PHILOSOPHY IN MEDICINE

Philosophy in Medicine

Conceptual and Ethical Issues in Medicine and Psychiatry

Charles M. Culver · Bernard Gert

New York Oxford
OXFORD UNIVERSITY PRESS
1982

Library of Congress Cataloging in Publication Data

Culver, Charles M.
Philosophy in medicine.

Bibliography: p.
Includes index.
1. Medicine—Philosophy. 2. Medical ethics.
3. Psychiatry—Philosophy. 4. Psychiatric ethics.
I. Gert, Bernard, 1934– . II. Title. [DNLM:
1. Ethics, Medical. 2. Philosophy, Medical. W 61
C968p]
R723.C84 1982 174'.2 81-14232
ISBN 0-19-502979-8 AACR2
ISBN 0-19-502908-1 (pbk.)

Printing (last digit): 9 8 7 6 5 4 3 2 1

Printed in the United States of America

To my mother, Virginia M. Culver,
and the memory of my father, Vernon A. Culver

To my wife, Esther Gert

Preface

This book is intended to demonstrate the value of philosophy in medicine. We hope it also demonstrates the value of close and continuing collaboration between philosophers and physicians. The two of us have been collaborating in research and teaching for many years and this book is one result of that collaboration. This book could not have been written by either of us working separately; it required the two of us working together, each teaching the other as we went along.

We were very fortunate to be in an institutional setting like Dartmouth, where the close physical proximity of College and Medical School facilitated cooperation. We were also fortunate that the administration of both schools fully supported our efforts.

Our work has profited enormously because we have colleagues in both the College and the Medical School who are also interested in interdisciplinary work in medicine and philosophy. Two of the chapters in this book (7 and 8) were developed for an institution-wide University Seminar on Philosophy and Psychiatry (now Philosophy and Medicine). Our colleagues in this seminar were an invaluable aid to our research, as well as producing many significant papers of their own.

Our research was aided by our being able to teach together. This joint teaching, in both the College and the Medical School, was supported by several grants. The two of us received a QUILL (Quality in Liberal Learning) grant from the American Association of Colleges and we benefited from an institution-wide grant from the Commonwealth Fund. Not only was our research aided by our teaching, we think that our teaching has benefited from our research. We have taught on a continuing basis: an undergraduate college course in the Philosophy of Medicine; an elective in Medical Ethics for first and second year medical students; a section of a required course for all second year medical students; a series of seminars for final year residents in psychiatry; and a monthly seminar for nursing clinical specialists at Mary Hitchcock Memorial Hospital. We found that the material in this book could be used successfully with all of these groups.

Chapter 4, "Maladies," was originally published in *The Hasting Center Report* as "Malady: A New Treatment of Disease." This article was written by us together with K. Danner Clouser of Hershey Medical School. We want to thank Dan, not only for his permission to use this article in our book, but also for helping us at the very beginning of our entry into the collaboration between medicine and philosophy. As he has done at so many other schools, Dan helped Dartmouth start its philosophy of medicine and medical ethics program.

Some of the chapters in this book were originally printed as articles. Chapters 7 and 8, "Paternalistic Behavior" and "The Justification of Paternalism," were published in *Philosophy and Public Affairs* and *Ethics* respectively, and though they have been reprinted in various anthologies, we have extensively revised both of them during preparation of this book. Chapter 9, "The Morality of Involuntary Hospitalization," though printed in a collection by Spicker, Healey, and Engelhardt, was originally intended for this book and so needed very little revision.

Chapter 10, "The Definition and Criterion of Death," was originally published in the *Annals of Internal Medicine*. This article was written by us together with James Bernat, a neurologist at the Dartmouth Medical School and the Veterans Administration Hospital in White River Junction, Vermont. We also want to thank Jim for his permission to use this article in our book, and for the continuing work we have done together on this topic. We think that Chapter 10 is an improvement over the original article, at least in part because of the work the three of us have done in writing a review of the President's Commission Report on the Definition of Death. We believe that our critique of this Report shows the relevance of careful philosophical analysis to public policy on medical matters.

The remaining five chapters were written specifically for this book. Chapter 1, "The Uses of Philosophy in Medicine and Psychiatry," does not just describe the uses of philosophy in medicine; it actually uses philosophical analysis in examining several matters of interest to physicians. Chapter 2, "Rationality and Irrationality," examines one of the most important concepts in medicine and philosophy and develops an original account of that concept, first proposed in *The Moral Rules*. Chapter 3, "Valid Consent and Competence," examines two issues which are among the most significant in medicine today. These issues have become a focus of attention not only for philosophers and physicians, but also for lawyers and all of those who are concerned with public policy. This chapter, like the previous one on Rationality and the following one on Maladies, demonstrates the close relationship between facts and values, and shows that reasoning about values can be as precise and objective as reasoning about facts.

Chapters 5 and 6, "Mental Maladies" and "Volitional Disabilities," are of special interest to those concerned with psychiatry. In both of these chapters we discuss aspects of the *Diagnostic and Statistical Manual of Mental Disorders, 3rd Edition* (*DSM-III*), published in 1980 by the American Psychiatric Association. We have great admiration for this work, and think that it deserves scholarly attention, not only from psychiatrists, but also from philosophers, especially philosophers of science. We think that close collaboration between philosophers and psychiatrists could make *DSM-IV* an even more valuable contribution to the development of scientific medicine. We hope that in these two chapters we have demonstrated some of the ways in which this collaboration would be fruitful.

Chapter 6, "Volitional Disabilities," is the development of a concept first put forward in 1967 by Timothy J. Duggan and Bernard Gert. This concept was used in 1979 by Gert and Duggan in discussing the traditional problem of Free Will. With some significant revisions, the first half of Chapter 6 is taken from this latter article. Then called the "Ability to Will" and now called "Volitional Ability," this concept, though developed to solve a purely philosophical problem, now seems to have considerable value in clarifying some issues in psychiatry. We are using this concept in further work on alcoholism and think that it may be of some practical value to physicians dealing with noncompliant patients. We are grateful to Tim Duggan for letting us make such extensive use of the earlier articles and for the help and encouragement that he has provided us in the writing of this entire book.

The topic of paternalism, which occupies Chapters 7 and 8, seems to us to lie at the heart of the doctor-patient relationship. We have found that discussion of this topic leads to almost every other topic in the area of doctor-patient relationships. Once one becomes clear about the nature of paternalistic behavior and can determine when it is and when it is not justified, many superficially unrelated issues become clearer. For example, we found that "The Morality of Involuntary Hospitalization," Chapter 9, turned out largely to be a special application of the issue of when paternalistic behavior is justified.

The topic of the final chapter, appropriately enough on "The Definition and Criterion of Death," has become a major issue in public policy. We hope that this chapter will contribute to that discussion. Though our original article was cited favorably by the President's Commission, we find ourselves in serious disagreement with the Uniform Determination of Death Act (UDDA) recommended by the Commission Report. In this chapter we incorporate some of the objections to this statute that we developed in our review of the Commission Report, to appear in *The Hastings Center Report*. We have put forward a modification of this statute, which meets the objections that we have raised, without, as far as we can see, raising any new objections of its own. Since the fault we find with the statute is primarily philosophical or conceptual, we think that this is another example of the use of philosophy in medicine.

We wish to acknowledge the assistance we have had in writing this book. One of us (Culver) is the recipient of a grant from the Ira W. De Camp Foundation and the other (Gert) has a Sustained Development Award from the National Endowment for the Humanities and the National Science Foundation, Grant number ISP-8018088 AO. Of course, all the views expressed in this book are the views of the two of us and do not necessarily reflect the views of the Foundations and agencies mentioned above.

We also wish to mention many colleagues who have been helpful to us in our efforts. No doubt we will miss some, but it is better to make the effort than to risk giving the impression that we have done everything on our own. First of all we must thank, again, James M. Bernat, K. Danner Clouser, and Timothy J. Duggan who actually helped to write the articles that were the basis of three chapters of this book. Then we must thank the members, past and present, of the University Seminar on Philosophy and Psychiatry, without whose stimulation we would probably never have begun on this venture: Bernard Bergen, K. Danner Clouser, Richard B.

Ferrell, Ronald M. Green, James Moor, Trevor Price, Joel Rudinow, Stanley Rosenberg, Raymond Sobel, Charles Solow, Gary Tucker, and Peter Whybrow.

We were fortunate to have the services of Harry Robinson, who not only was a flawless typist, but who demonstrated to us the value of using a computer in preparing a manuscript. By typing our manuscript into a computer he allowed us to make many more changes than we would have made otherwise. A very large number of those changes were required because of the work of Huntington Terrell of Colgate University, who read the entire manuscript and provided us with a detailed list of both substantive and stylistic suggestions. Even those few suggestions we did not accept forced us to say more clearly why we proposed the position that we did. We are also deeply indebted to Jeffrey House of Oxford University Press, who not only encouraged us but also provided many substantive suggestions for improving the manuscript.

Hanover, N.H. C.M.C.
November, 1981 B.G.

Contents

PHILOSOPHY IN MEDICINE

1

The Uses of Philosophy in Medicine and Psychiatry

Is alcoholism a disease? Are there any mental illnesses? Can there be rational suicide? Can there be some schizophrenics who respond well to lithium and not to the major tranquilizers? It may seem obvious that these first three questions are not merely empirical, that they involve philosophical or conceptual issues. But the fourth question may seem to be a straightforward empirical matter. However, it is not, for some may hold that a response to lithium is diagnostic of manic-depressive illness no matter how much the behavior may seem to be that of a schizophrenic, while others may hold that it is the behavior, history, and other diagnostic tests that determine whether someone is suffering from schizophrenia or manic-depressive illness. Thus two doctors may agree on all the facts of a particular case and even the prognosis but still differ as to the correct description of the patient's illness. When this happens, it is most likely that conceptual issues are involved, and that philosophical analysis is needed in order to resolve the dispute. Thus, though medicine and psychiatry are permeated with philosophical problems, it is not always obvious that this is so; philosophical problems can masquerade as empirical problems.

Empirical problems are those that can, at least in principle, be solved by the collection of appropriate data and the analysis of those data by relevant statistical techniques. But sometimes we have all the facts, and there are no further facts that would resolve an issue. In these cases, something different must be done. We must begin to examine the concepts involved. For example, when Thomas Szasz (1961) asserts that mental illness is a myth, he is not simply making an empirical claim that can be conclusively verified or falsified by the collection of new information. Szasz is aware of the same facts as other psychiatrists, who have no doubt that mental illnesses are anything but mythical. Both Szasz and those who believe in the reality of mental illness know that there are people who behave in ways that show they have a distorted view of reality, and that others act in ways that are very harmful to themselves; they also both know that some people find it almost impossible to enter elevators and that others wash their hands so often that they become cracked and sore. They know too that some of these people respond to lithium and the various antipsychotic and antidepressive drugs, while others respond to various psychotherapies. What then are they in dispute about? They disagree about the proper use of the term "illness." Szasz holds that illness or disease must be related to some identifiable malfunction in the body, so that all illnesses are, by definition, bodily illnesses. Others find this to be an inadequate account of the nature of illness; they hold that illness should be defined in terms of symptoms and that someone suffering from certain symptoms is ill regardless of whether or not there is any identifiable part of the body that is malfunctioning. It should be clear that this dispute cannot be settled solely by reference to the facts. What is needed is an analysis of the concept of illness. It may seem odd, but there have been few serious attempts to do this in the history of medicine until quite recently.

Philosophers have as one of their primary tasks the analysis of concepts, and in recent years philosophers have come to realize that medicine, especially psychiatry, employs many interesting concepts that are in need of analysis. The concepts of illness in general and mental illness in particular are two important examples, and we will analyze these in detail in Chapters 4 and 5. Two other concepts of great importance in medicine and psychiatry are those of rationality and competence. Can there be a rational suicide? Can someone suffering from a major psychosis be competent to consent to a given treatment? What facts can decide this issue? It should be clear that before these questions can be answered, we must be clear about the concepts of rationality and competence. Philosophers try to clarify such concepts by seeing how they are actually used and what role they play in the practice of medicine and psychiatry. We will analyze the concepts of rationality and competence in Chapters 2 and 3.

Concepts are intimately related to words; we shall attempt to clarify concepts by examining the meaning of the words that are related to them. To do this, one must look at the way these words are used in their natural setting. With most words this setting is ordinary life, but with some words one must see how they are used in a more technical setting, for example, in psychiatric practice. It sometimes happens that words (e.g., "depression") are used in related though distinct ways in ordinary life and in a technical setting, and then an attempt must be made to clarify these distinct uses. In this first chapter, we will give examples of some of the methods of philosophical analysis which can be applied to conceptual problems in medicine and psychiatry and which will be used throughout this book.

Increasing our understanding of the meaning of words and concepts

First, it is important to distinguish between knowing the meaning of a word and being able to give a definition of it. A person knows the meaning of a word if he is able to use it correctly and can respond correctly to other persons' use of it. All of us know the meanings of the words we use regularly in ordinary everyday life and in our standard medical practice. But in order to increase our understanding of such words, it is very useful to generate explicit definitions of them. These explicit definitions, if they are adequate, will give the criteria for the use of a word in our language. We will illustrate these notions by examining an inadequate definition of paternalism.

Paternalism

"Paternalism" is a concept of great importance in medical practice. One sees very frequent allusions to it in discussions of the doctor-patient relationship. Participants in these discussions seem to have a fairly good sense of the meaning of the term, but occasionally there is disagreement over whether or not a given act is parternalistic. *Webster's New Collegiate Dictionary* offers the following definition of paternalism: "A system under which an authority treats those under its control in a fatherly way, especially in regulating their conduct and supplying their needs." Extrapolating from this definition, we might define "paternalistic behavior" as acting in a fatherly way, especially in regulating the conduct and supplying the needs of those under one's authority. But this definition is obviously too broad. Fathers act in all sorts of ways in regulating the conduct of their children and in supplying their needs. If a father offers a child a dollar for raking the lawn, is that paternalistic? It is a fatherly way of regulating conduct, but we would not usually think of it as paternalistic. When we think of supplying needs, the definition is even more clearly inadequate. Buying one's hungry child a hamburger might be a common fatherly act, but it does not seem remotely connected to paternalism. Thus, acting fatherly obviously includes much behavior that is not paternalistic. Further, acting fatherly excludes as paternalistic behavior much behavior that should be included. Indeed, overprotective behavior, which is not normally thought of as fatherly, but rather as motherly, is often clearly paternalistic. The author of the definition might rejoin by saying that "fatherly" was not meant literally and that when mothers are overprotective they are acting fatherly. But that answer would leave us completely in the dark about what "fatherly" means in the definition except that it is intended as a synonym for "paternalistic"; thus, we would be left with a totally circular definition.

This discussion shows how we can test a proposed definition of a concept without even having a definition of our own. We can test the dictionary's definition because we do know the meaning of "paternalistic" in the sense that we know whether we would or wouldn't apply the term to a situation like buying a child a hamburger. Thus we know how to use the word in our language. And just as one may be able to swim without being able to describe what one is doing when swimming, so one may be able to use a word correctly (know its meaning) without being able to give a description of its use (know its definition). One problem with inadequate definitions is that they sometimes lead one to distort the use of the term in order to conform to the definition. For example, suppose one incorrectly defines paternalistic behavior as interfering with someone for his own good. Then, since this definition does not say "someone else", one notices that one can, consistent with this definition, act paternalistically toward oneself (see Husak, 1981). One may therefore begin to term prudential behavior designed to prevent oneself from giving in to future temptation as paternalistic behavior. This distortion of the concept, because of an inadequate definition, may make it more difficult to appreciate an adequate definition.

To return to the dictionary definition of paternalism: another way in which it is inadequate is that it misses an important evaluative feature of paternalism. To label a behavior as "paternalistic" is generally taken as a criticism of it. If one attempts to explain this feature, one discovers that paternalistic behavior has this pejorative connotation because it always involves doing something to someone that requires moral justification. We express this point by saying that it always involves the violation of a moral rule, that is, that every act of paternalism involves doing something like causing someone pain, deceiving someone, depriving someone of freedom, and so on. We will not explore this matter in detail here; instead, in Chapter 7 we will discuss at length all of the features of what we believe to be an adequate definition of paternalistic behavior. However, we mention here that a moral rule violation is part of paternalistic behavior because it illustrates how careful philosophical analysis can increase our understanding of a concept. Some significant work has been done by philosophers on the concept of moral rules, including the question of how to determine justified exceptions to them. Therefore, connecting the concept of paternalistic behavior to the concept of the violation of a moral rule not only explains why calling behavior paternalistic is often a criticism of it, it also enables us to deal with the question of when paternalistic behavior might be justified. Thus a body of philosophical work developed elsewhere can be brought to bear on the subject of paternalism and greatly deepen our understanding of it.

Freud's analysis of the concepts of "mental processes" and "sexuality"

One of the most interesting areas in which the analysis of concepts has played an important role is in psychoanalysis. Part of the importance of Freud's legacy lies in his successful attempt to provide a better analysis of the concepts of "mental processes" and "sexuality" than had previously been offered. The philosophical nature of his work is partly concealed because Freud called the views that he put forward "hypotheses." He said, "Two of the hypotheses of psychoanalysis are an insult to the entire world and have earned its dislike" (1966, p. 21). These two hypotheses concerned mental processes and sexuality. We shall examine these hypotheses with a dual purpose: to achieve a clearer understanding of Freud's views on these matters and to show the importance of conceptual analysis and the careful examination of the meaning of words.

According to Freud, "The first of these unpopular assertions made by psychoanalysis declares that mental processes are in themselves unconscious and that of all mental life it is only certain individual acts and portions that are conscious" (1966, p. 21). We must first be clear what this hypothesis means. Is declaring that some mental processes are unconscious a simple empirical assertion? That is, is it quite clear what counts as a mental process and that we need only see whether these processes are sometimes unconscious? Freud himself did not think this to be the case. He said, "we are in the habit of identifying what is psychical with what is conscious. We look upon consciousness as nothing more or less than the *defining* characteristic of the psychical" (1966, p. 21). These remarks show that Freud realized that he was challenging our ordinary understanding of the concept of mental or psychical processes. It was clear to him that if consciousness was taken as a defining characteristic of mental processes, then there could be no unconscious mental processes. Thus, Freud tries to show that defining mental processes in this way is inadequate.

He starts to do this by calling attention to our ordinary use of the phrase "mental process"; he offers an ostensive definition of the phrase. Strictly speaking, an ostensive definition involves teaching the meaning of a word by pointing to those things that the word refers to, for example, teaching the word "chair" by pointing to many different chairs. But one can also give an ostensive definition by listing examples, and this is what Freud does. He says that psychoanalysis "defines what is mental as processes such as feeling, thinking and willing" (1966, p. 22). Everyone would agree that feeling, thinking, and willing are mental processes, but we are still faced with the problem of whether there can be unconscious feeling, thinking, and willing. Freud claimed that psychoanalysis "is obliged to maintain that there is unconscious thinking and unapprehended willing" (1966, p. 22), but those who use consciousness to define mental processes will regard it as nonsense to allow for thinking and willing of which we are unaware.

Freud thinks that such people are like those who regarded whiteness as a defining characteristic of swans; they did so because they were unaware of those black birds in Australia which were biologically identical to swans in every way except color. Once one became aware of these birds, it would be pointless to continue using whiteness as a defining characteristic of swans; to do so would be to use the word "swan" in a way that was inconsistent with all sorts of similar words, for example, "duck," in which color was not a defining characteristic. Freud is making a similar point in the following statement: "To most people who have been educated in philosophy the idea of anything psychical which is not also conscious is so inconceivable that it seems to them absurd and refutable simply by logic. I believe this is because they have never studied the relevant phenomena of hypnosis and dreams, which—quite apart from pathological manifestations—necessitate this view" (1960, p. 3).

What Freud is doing is pointing out that certain kinds of behavior require us to say that the person is thinking or willing, even though on some of these occasions the person himself may sincerely and correctly deny that he is engaged in any conscious mental process. He provides many, many examples of this kind of behavior; the various parapraxes, including slips of the tongue, which we now often refer to as "Freudian slips"; neurotic symptoms, especially obsessional acts; dreams; and, most importantly, hypnosis. Freud himself regarded hypnosis as the incontrovertible proof of unconscious mental processes. It is hypnosis which makes everyone admit that mental processes can be unconscious and thus forces them to consider it an empirical possibility that there are other examples of such processes. This belief comes out clearly in the following question and answer. "Where, then, in what field, can it be that proof has been found that there is knowledge of which the person concerned nevertheless knows nothing, as we are proposing to assume of dreamers?" (1966, p. 102). Freud's answer is, "The proof was found in the field of hypnotic phenomena" (1966, p. 103). Freud then recounts his witnessing of an experiment by Bernheim in which a man was hypnotized, made to experience all kinds of things, and when awakened appeared to know nothing of what had happened during his hypnotic sleep, but under considerable pressure finally became conscious of it. Freud concludes, "Since, however, he knew afterwards what had happened and had learnt nothing about it from anyone else in the interval, we are justified in concluding that he had known it earlier as well. It was merely inaccessible to him; he did not know that he knew it and thought he did not know it" (1966, p. 103).

Freud uses this information about hypnosis to argue for the plausibility of unconscious mental processes. He says, "If a person thinks he knows nothing of experiences the memory of which he nevertheless has within him, it is no longer so improbable that he knows nothing of other mental processes within him" (1966, p. 103). And later, he uses hypnosis again to make the case for the existence of unconscious mental processes in obsessional action. After discussing a patient in whom he claims unconscious mental processes were at work he continues, "She behaved in precisely the same way as a hypnotized subject whom Bernheim had ordered to open an umbrella in the hospital ward after he woke up. The man carried out the instruction when he was awake, but he could produce no motive for his action. It is a state of affairs of this sort that we have before our eyes when we speak of the existence of *unconscious mental processes*" (1966, p. 277). Why is Freud so confident that this acting on a posthypnotic suggestion shows that there are unconscious mental processes? What can Freud mean by a "mental process" now that it is not to be defined as some process of which we are conscious?

Let us go back to Freud's ostensive definition of mental processes as feeling, thinking, and willing. People explain some of their behavior by invoking these mental processes, e.g., they say that they went to the dictionary because they were thinking about the spelling of a certain word. Also, sometimes we explain the behavior of others by invoking these same kinds of mental processes, and the person whose behavior we are explaining generally confirms our explanation. For example, someone is studying, and he suddenly stops, rushes over to the phone, and calls to wish his mother a happy birthday. We say that he suddenly remembered that it was his mother's birthday and that he was supposed to call her. After the call, we ask him and he confirms our hypothesis. But in some cases, the person does not confirm our hypothesis. he is not aware of any mental process which explains his behavior. What are we to say in such cases? Freud believes that, if our evidence is good enough, we should say that there were indeed mental processes responsible for his behavior, only the person was not aware of them; they were unconscious mental processes. Hypnosis provides cases in which the evidence is so good that it becomes virtually impossible to deny that what explains the person's behavior is a mental process similar in every respect to conscious mental processes except that it is unconscious.

Consider again the case of the subject who opened the umbrella five minutes after he was awakened. We know that the only plausible explanation of his behavior is that he "remembered" the instructions he was given under hypnosis and acted because of them, even though he was not conscious of this. The whole sequence of events differs from the workings of ordinary mental processes only in one respect: the person in whom the mental processes are operating is unaware of their operation. To deny that there were unconscious mental processes working would be like denying that the black swans of Australia were really swans; it would reject the overwhelming similarities between two kinds of events (or birds) because of one prominent but superficial characteristic, consciousness (or color). But just as asserting that black swans are swans requires one to have a concept of swan in which color is not a defining characteristic, so asserting that unconscious mental processes are mental processes requires having a concept of mental processes in which consciousness is not a defining property.

The biologist does have such a concept of swan. It is related to the wider notion of a species, and it fits into an integrated theoretical framework. Freud does have such a concept of mental process, and it is related to a wider theoretical framework for explaining behavior. For Freud, mental processes are primarily those processes which are necessary to explain those aspects of human behavior not explainable by the laws of physics, chemistry, or biology, for example, all learned or culturally determined aspects of our behavior. Freud has noted the actual use we make of mental process terms: "I was so nice to him because I was *thinking* of all of the trouble he had just been through." "I hit him because I *felt* so angry." We use mental process terms to explain our behavior. So, Freud simply takes mental processes to be those processes that can explain our learned or culturally determined behavior. Even more significant, as with hypnosis, we may not be aware of a mental process at a particular time, but can be made to become aware of it at some later time. This seems conclusive evidence that our behavior is explained the same way whether we are aware of it at the time or not. If it is a mental process when we are conscious of it when it happens, why is it not a mental process if we become aware of it only a minute or two later, as in some parapraxes, or an hour or so later as in hypnosis, or a week, month, or year later, as with the unconscious mental processes that explain some neurotic behavior?

Once we define mental processes as those which explain certain types of behavior, it becomes quite clear that these processes can operate without our being conscious of them. This dramatically expands the province of psychology; behavior can be a reliable sign of mental processes at work without or even contrary to introspective reports by the agent. Thus Freud is certainly correct in his assessment of the importance of his new view of the nature of mental processes. He says, "The question whether we are to make the psychical coincide with the conscious or make it extend further sounds like an empty dispute about words; yet I can assure you that the hypothesis of there being unconscious mental processes paves the way to a decisive new orientation in the world and in science" (1966, p. 22).

But it would be incorrect to think that Freud has created a totally new view of mental processes. Rather, he has explicitly and systematically stated a view which has been recognized by most people implicitly, and by many great writers explicitly. We all believe that "actions speak louder than words," that what a person does often reveals his mind better than what he says. Shakespeare's plays are filled with characters who act out of love without their knowing it, though everyone else does. In short, Freud has done what all great philosophers have done: he has reminded us of what we all know but did not know that we knew, because we were confused by our own mistaken thoughts about the language. We all know that we sometimes act because of thoughts and feelings we are unaware of at the time of acting, but we overlook this because we mistakenly take consciousness as a necessary characteristic of mental processes. By forcing us to look at how we actually use the concept of mental process, especially in the dramatic case of hypnosis, Freud made us realize that mental processes do not need to be conscious.

Freud's second hypothesis "is an assertion that instinctual impluses which can only be described as sexual, both in the narrower and wider sense of the word, play an extremely large and never hitherto appreciated part in the causation of nervous and mental diseases" (1966, p. 22). This hypothesis is somewhat more complex than the first, which concerned mental processes. On our interpretation, that hypothesis was almost entirely concerned with the analysis of the concept of mental process; we take this hypothesis as not only an analysis of the concept of sexuality but also as an empirical assertion about the causation of nervous or mental diseases. Of course, these two features of the hypothesis are closely related, because a change in the concept of sexuality is necessary before the empirical assertion can even become plausible, but in what follows we shall be concerned only with the analysis of the concept of sexuality.

Freud himself draws a parallel between his investigation of the concept of sexuality and his account of the psychical. He says, "Whereas for most people 'conscious' and 'psychical' are the same, we have been obliged to extend the concept of 'psychical' and to recognize something 'psychical' that is not 'conscious.' And in just the same way, whereas other people declare that 'sexual' and 'connected with reproduction' are identical, we cannot avoid postulating something 'sexual' that has nothing to do with reproduction" (1966, p. 22). While hypnosis played the crucial role in forcing a change in our concept of mental processes, "sexual perversions" play a similar role in forcing a change in our concept of sexuality. (Freud used the now antiquated term "sexual perversions" for those conditions currently labeled as "paraphilias" by the American Psychiatric Association's *Diagnostic and Statistical Manual of Mental Disorders, 3rd ed.* [1980]. We will discuss these conditions at some length in Chapter 5.) In talking about the various "perversions", such as homosexuality, oral and anal sex, and fetishism, Freud says, "As is already shown by the name by which they are universally known, they are unquestionably sexual. Whether they are described as indications of degeneracy or in any other way, no one has yet had the courage to class them as anything but phenomena of sexual life" (1966, p. 320). Freud regards these sexual perversions as decisive in showing that sexuality cannot be regarded as being essentially connected with reproduction.

But Freud acknowledges that sexuality may not clearly be separated from the genitals. He acknowledges that, at least in adults, "What makes the activities of perverts so unmistakably sexual in spite of all the strangeness of its [sic] objects and aims is the fact that as a rule an act of perverse satisfaction nevertheless ends in complete orgasm and the voidance of genital products" (1966, p. 321). Freud thus seems to define sexual activity as any activity which, as a rule, ends in genital satisfaction. We hold that in his hypothesis, when he talks of impulses being sexual "both in the narrower and wider sense of the word," the narrower sense is that sense in which what is sexual is related to genital satisfaction. He acknowledges that determining what is sexual by its relation to genital activity is a strong position, but he states that his own position is different in that he believes that pleasure derived from other organs, for example from the mouth by kissing, is sexual as well (1966, p. 321). We suggest that it is activity which results in pleasure in organs other than the genitals that Freud means by "sexual" in its wider sense. One could say that his wider sense of "sexual" is simply equivalent to the ordinary use of "sensual," but though we think this correct, it fails to do justice to Freud's view that in children the natural development is to go from taking (bodily) pleasure from stimulation of various organs of their body to taking (bodily) pleasure primarily from stimulation of their genital organs.

By showing that sexuality has more to do with genital pleasure than with reproduction, Freud has opened up the possibility that taking pleasure in organs of the body not connected with reproduction might also properly be considered sexual. This is parallel to what he did by showing that the psychical might not be conscious; he opened up the possibility of asking new questions. We can now at least ask whether some of the activities of prepubertal children are sexual in character, and whether certain kinds of prepubertal "sexual" experiences may in fact play a role in the development of adult perversions. Without the insights derived from Freud's conceptual analysis (for example, if sexuality were defined solely in terms of reproduction), these questions would have been considered nonsensical. We take no stand on Freud's theory of the sexual life of children, but we do hold that by explicitly connecting sexuality with pleasure derived from stimulation of various parts of the body, rather than with reproduction, Freud has helped us to understand the concept of sexuality far better than we did.

Making conceptual distinctions

Sometimes, determining how a word is used in ordinary language or in technical language is a reasonably straightforward task, consisting mainly of a careful investigation of the criteria for its use. However, on other occasions, certain complications come to light as a result of one's analysis. For example, a word may prove to have two (or more) related but distinct uses that users of the language are unaware of, or only vaguely aware of, until that fact is explicitly pointed out. In such cases, it is frequently illuminating to identify and clarify these distinct uses. In addition, in order to avoid the conceptual quandaries and ambiguities that can result from a term's having two or more distinct meanings, sometimes it is useful to suggest that henceforth, at least in technical usage, a term should always be employed to convey just one specific meaning. A classical example of the usefulness of making conceptual distinctions is provided by William James with regard to the concept of "going around."

James' analysis of going around a squirrel

James wrote as follows (1955, p. 41):

> Some years ago, being with a camping party in the mountains, I returned from a solitary ramble to find everyone engaged in a ferocious metaphysical dispute. The corpus of the dispute was a squirrel—a live squirrel supposed to be clinging to one side of a tree trunk; while over against the tree's opposite side a human being was imagined to stand. This human witness tries to get sight of the squirrel by moving rapidly around the tree, but no matter how fast he goes, the squirrel moves as fast in the opposite direction, and always keeps the tree between himself and the man, so that never a glimpse of him is caught. The resultant metaphysical problem now is this: Does the man go round the squirrel or not? He goes round the tree, sure enough, and the squirrel is on the tree; but does he go round the squirrel? In the unlimited leisure of the wilderness discussion had been worn

threadbare. Everyone had taken sides and was obstinate; and the numbers on both sides were even. Each side, when I appeared, therefore appealed to me to make it a majority. Mindful of the scholastic adage that whenever you meet a contradiction you must make a distinction, I immediately sought and found one, as follows: "Which party is right," I said, "depends on what you practically mean by 'going round' the squirrel. If you mean passing from the north of him to the east, then to the south, then to the west, and then to the north of him again, obviously the man does go round him, for he occupies these successive positions. But if on the contrary you mean being first in front of him, then on the right of him, then behind him, then on his left, and finally in front again, it is quite obvious that the man fails to go round him for by compensating movements the squirrel makes, he keeps his belly turned towards the man all the time, and his back turned away. Make the distinction, and there is no occasion for any further dispute. You are both right and wrong, according as you conceive the verb 'to go round' in one practical fashion or the other." Although one or two of the hotter disputants called my speech a shuffling evasion, saying they wanted no quibbling nor scholastic hair-splitting, but just plain English "round," the majority seemed to think that the distinction has assuaged the dispute.

In ordinary life, we usually have no occasion for distinguishing the two different ways of going round something, and on those rare occasions that we do need such a distinction, we can make it in the way that James did. We usually don't make the distinction because when we go round in one way we almost always also go round in the other, so we do not realize the two separate senses of the phrase. Once we become aware of the two senses, we realize that the phrase "going round" is not sufficiently precise to determine whether one or the other or both of these senses is intended. The only way to solve this dispute is as James did, to make a distinction and to provide two separate definitions of the two senses of "going round." Using these two precise definitions, we can now say that it is true that the man went round the squirrel in one of these senses and not in the other. In James' case, both senses of "going round" seem of equal practical value, and there seems no good reason for assigning the phrase "going round" to one and making up a new phrase for the other. Furthermore, nothing else turns on whether we decide that the man went round the squirrel or not. If there were some complex ritual in which going round a squirrel was important, then it is very likely that we would examine the ritual and see which of these senses fit in more closely with what seemed to be the point of the ritual. We could then say that the important sense (perhaps going front-left-back-right of) is what we would henceforth mean by "going round," and the alternative sense should be called "encircling."

James has provided a model for what we call making a conceptual distinction. We take a term or phrase which has two or more distinct criteria for its application, which normally occur together. We find a case in which they do not go together and then decide that one of the criteria is more important than the other and that it alone picks out the essential feature; the other feature is now demoted to a merely contingent or factual matter. We acknowledge that such a procedure is not an accurate account of ordinary usage, but ordinary usage is sometimes misleading because it presupposes a constant correlation between two features which turn out not always to occur together.

All of this may seem interesting but rather abstract and most likely of no practical value. But this is not true: providing precise definitions can be of great practical import, as we shall attempt to show in the following chapters. Here we shall anticipate our discussion of one of the more dramatic examples.

The definition of death

Before the development of modern resuscitative and life-support technology, the question of when someone had died was rarely a matter of dispute. The usual precipitating cause of death was the cessation of cardiopulmonary function. This was shortly followed by the permanent loss of function of the whole brain, which then led to the permanent loss of function of all organ systems throughout the body. Once this sequence of events had been set in motion, it proceeded rapidly in a regular and ineluctable fashion. The loss of spontaneous cardiac and pulmonary function was quickly and inevitably followed by permanent coma and then by permanent nonfunctioning of the entire organism; thus it did not seem to matter what definition of death was used.

However, modern technology, by making it possible to resuscitate people after several minutes of loss of adequate cardiac function, has created the following puzzle. Some of these individuals have a permanently destroyed neocortex but an at least partially preserved brainstem. Thus they have a flat electroencephalogram and are totally and permanently comatose. However, they have normal spontaneous respiration and heartbeat. With vigorous medical and nursing care, some of these patients can be preserved in this state for months (see Brierley et al., 1971, for a full description of such patients). Should patients in this condition be considered alive or dead?

When one first thinks about this question, a kind of cognitive puzzle may occur similar to that in James' example of going round the squirrel. One might contend that anyone whose neocortex is completely destroyed and who thus is not only unable to move at all but is totally and permanently comatose really ought to be considered dead. On the other hand, it seems extremely odd to think of a patient as dead who is lying in bed, unattached to any life-support apparatus, with normal spontaneous heartbeat and breathing.

As in James' example, one can better understand this cognitive puzzle if a conceptual distinction is made. There are two ways in which death might be defined. In the first, death means the permanent cessation of functioning of the organism as a whole. In the second, death means the permanent loss of what is essential to personhood. The patients described above would be dead if we used the second definition, for they have ceased forever to be "persons" as we usually use that term. They will never again be conscious, show evidence of any mental processes, or be able to interact with others. However, they are not dead if we use the first definition because, considered as organisms, they have not ceased to function as a whole: Their cardiopulmonary functioning is still intact (as are other physiological processes, since at least some of their brain stem is preserved).

Which definition of death should we choose? One can see why this is a modern problem: Before the advent of modern therapeutics, patients almost never permanently lost what was essential to their personhood without at the same time or very quickly thereafter undergoing a permanent cessation of functioning of their organism as a whole (just as the two senses of "going round" almost always occur together). Unlike the "going round" example, here one's choice between the two definitions is a matter of great practical importance, since patients like those described above will be declared alive or dead depending on which definition is chosen. We will give our reasons for choosing one of these definitions in Chapter 10.

As a final example of philosophical analysis in medicine, we will begin our discussion of two very important concepts: incompetence and irrationality.

The distinction between incompetence and irrationality

Consider a forty-six-year-old man, Mr. E, who is suffering a severe episode of depression and wishes to die. He has suffered from two similar episodes during the past ten years and on both occasions his depression, while necessitating in-patient care, responded fairly readily to antidepressant medication and supportive psychological therapy. After each of these depressive episodes, he returned to his usual level of functioning, including a job and a family life which he enjoyed.

He has now become depressed again and, as in his two earlier episodes, there is no apparent precipitating cause. He states that he wants to die but gives no clear, consistent reason other than his intense dejection and feelings of worthlessness. He is admitted to the psychiatric in-patient service at his wife's urging. He enters voluntarily but reluctantly and refuses either to take medications or to accept electroconvulsive therapy because of his stated wish to die. Over the next two weeks, his physical condition begins to deteriorate as a consequence of his anorexia and weight loss. The ward staff come to realize that Mr. E may indeed die unless they, perhaps with the sanction of his wife and the local probate court, begin to treat him without his consent.

When psychiatrists and others discuss patients like Mr. E, they are apt to label them with terms like "incompetent" and "irrational." These two terms are frequently used interchangeably and confusingly. It is generally overlooked that "irrational" is used primarily to characterize actions, while "incompetent" is used primarily to characterize persons, though the characterization is related to the person's ability to do certain kinds of actions.

A patient like Mr. E, who refuses a treatment which his doctors believe is strongly indicated, may be incorrectly labeled as incompetent solely because of what may be correctly characterized as an irrational decision. It is usually not recognized that if someone is labeled incompetent on the basis of an irrational refusal, the logic of incompetence requires that he also be incompetent to give consent. Whereas if it is Mr. E's refusal that is irrational, but he is regarded as competent, then were he to change his mind and give consent, no problem is created about whether he is competent to do so.

Distinguishing between "incompetent" and "irrational" is not sufficient to explain the meaning of either of these terms. "Incompetent" and "irrational" each have a complex use in our language, and it is still necessary to clarify these uses and to make explicit the criteria we employ when we correctly use either of these terms. We will consider these concepts in detail in the two succeeding chapters.

References

American Psychiatric Association. *Diagnostic and Statistical Manual of Mental Disorders, 3rd ed.* Washington, D.C.: American Psychiatric Association, 1980.

Brierley, J. B., Adams, J. H., Graham, D. I., and Simpson, J. A. Neocortical death after cardiac arrest. *Lancet*, 1971, *2*, 560–565.

Freud, Sigmund. *The Ego and the Id.* New York: Norton, 1960.

Freud, Sigmund. *Introductory Lectures on Psychoanalysis.* New York: Norton, 1966.

Husak, Douglas N. Paternalism and autonomy. *Philosophy and Public Affairs*, 1981, *10*, 27–46.

James, William. *Pragmatism.* New York: World, 1955.

Szasz, Thomas. *The Myth of Mental Illness.* New York: Hoeber-Harper, 1961.

2

Rationality and Irrationality

Introduction

"Rationality" and "irrationality" are among the most important concepts in both psychiatry and philosophy. Yet psychiatrists have generally not presented any explicit account of them and have often not distinguished these concepts from related ones like "mental health" and "mental illness." Further, there has been almost no attempt by philosophers to take psychiatric phenomena into account when formulating their own analyses of these concepts. We believe that this oversight has resulted in a distortion of these concepts, and in this chapter we shall attempt to remedy this oversight and provide an adequate analysis of the common sense or ordinary understanding of the concepts of rationality and irrationality.

Though "rational" and "irrational" are used in several distinct ways, all theories of rationality and irrationality agree that to label something as irrational is to express an unfavorable attitude toward it. Thus to say that a proposed action is irrational is to advocate that it not be done; an irrational belief or desire is not to be acted on; and one should even avoid having irrational beliefs and desires.

It is not quite so clear what is entailed by labeling something as rational. Some philosophers have claimed that to label something as rational is to express a favorable attitude toward it, so that to call a proposed action rational is to advocate doing it and to call beliefs and desires rational is to advocate holding them. But this seems to be far stronger than our ordinary understanding of the concept, for almost all agree that there can be two incompatible actions that are both rational. For example, there may be two rational treatment plans for depression and, though a physician prefers one, he need not consider the patient irrational for choosing the other. Further, we are aware that some actions are rational, in the ordinary sense of that term, and yet immoral, and so we would not advocate carrying them out even though we agree they are rational. However, to label an action as irrational is always to advocate that it not be done.

This asymmetry between rationality and irrationality, with irrationality having the clearer normative force, leads us to treat irrationality as more basic than rationality, and to define rationality in terms of irrationality. In this way, we can distinguish two classes of rationality: (1) that which it would be irrational not to do (or believe, or desire), and (2) that which it would not be irrational not to do (or believe, or desire). The former class of rational actions (or beliefs, or desires) we call rationally *required*, the latter class rationally *allowed*. That there are two classes of rational actions, beliefs, and desires enables us to understand why to call an action rational is not necessarily to advocate that it be done, for it may be only rationally allowed. However, this also explains why calling something rational sometimes has a stronger force, for one may be talking about something rationally required. On our account, whatever is not irrational is rational, either rationally allowed or rationally required.

We shall now consider a case of planned suicide and one of attempted suicide. We believe that suicide in the first case is rationally allowed, but that suicide in the second case would be irrational, or rationally prohibited. We shall use these cases in evaluating the two theories of rationality that we shall discuss prior to presenting our own account of irrationality.

Case 2-1.

Miss A is a 52-year-old woman with advanced and increasingly painful pancreatic cancer. She has been informed by her physician that her life expectancy is no more than two to three months and that little more than palliation can be offered to her. She has no close living relatives and few obligations. She can see no point in enduring several months of pain before experiencing her certain death. She has access to a lethal quantity of barbiturates and plans to take them after setting a few personal affairs in order.

Case 2-2.

Mr. K is a 42-year-old man with a history of impulsive behavior whose wife had confessed, earlier in the evening, that she was having an affair with one of his colleagues. He had become acutely agitated and depressed and, after several hours of mounting tension, told his wife that he was going to kill himself so "you can have all the lovers you want." There were guns in the house and when he left the room to get one, she called the police.

Most people feel very differently about these two cases. Miss A's plan to die does not seem irrational in her circumstances, though of course if she chose to live out her remaining months as best she could, that decision would not seem irrational either. However, Mr. K's suicide attempt does seem irrational since he appears to have no adequate reason for killing himself.

Past definitions of rationality

Rationality as the holding of true beliefs (*the intellectualist view*)

In this view, "rationality" does not apply primarily to actions but to beliefs, and an action is regarded as rational if and only if it is based on true beliefs. According to this definition, originally formulated by David Hume, one determines the rationality or irrationality of an action by looking not at the action but rather at the beliefs of the actor. If all of the relevant beliefs are true, that is, if the person knows both what he is doing and the probable consequences of doing it, then his action is rational, no matter what he does or why. This view of rationality can be taken as equivalent to saying that a person is acting rationally if he has an undistorted view of reality, and irrationally if he has a distorted view. An undistorted view is taken to mean having true beliefs, while a distorted view means having false beliefs. Taken literally, this view is clearly inadequate, for it would make every action based on false beliefs irrational, no matter how well the false belief is supported by the evidence. But the view can be modified to take account of this flaw; we shall not go into the various modifications because we believe that the view is basically incorrect.

The basic error of this view is that it confines the concept of rationality to the intellect and does not allow for irrational desires. To put the matter in more psychiatric terms, it allows only actions based on delusions to be labeled as irrational but not actions based on the desires that may accompany mood disorders. Thus, according to this view, someone like Mr. K, who is deeply agitated and depressed and desires to kill himself, is acting rationally when he attempted suicide if his action is not based on false beliefs. A defender of the intellectualist position might try to counter this objection by claiming that Mr. K does not realize his feelings are temporary or that a severely depressed person does not know that his depression can be cured; therefore, since his suicide attempt depends on such a false belief, or rather the absence of relevant true beliefs, his action can be counted as irrational. But a depressed person may have been depressed before and been cured, and thus may know that he can be cured yet still desire to die. The defender can then counter that the depressed person does not "really" know that he can be cured; he does not "genuinely" appreciate the possibility of a future when he is not depressed.

There is no final satisfactory resolution to this dispute; defenders of the intellectualist view of rationality can always insist that there must be some intellectual failing behind every irrational act. Even if one presented clear evidence that a person knew beforehand the complete uselessness of a self-harming act, and yet did it anyway, for example, as a result of extreme anger, the defender of the intellectualist view could say that the anger made the person temporarily forget. But all of these defenses of the intellectual theory indicate that it is just that, a theory; it is not a description of what we mean by an irrational action. It is not because we know that false beliefs are involved that we call an action irrational; rather, when we encounter acts of a certain kind, for example, some attempts at suicide or self-mutilation, we regard them as irrational. Then, if we defend the intellectualist view, we try to find the false beliefs that gave rise to that act.

There is one positive point in favor of the intellectualist view: acts that seem irrational, such as suicide or self-mutilation, can come to be viewed as rational if the person attempting it, like Miss A in Case 2-1, can be shown to have a nondelusional belief that she is suffering from an extremely painful terminal illness. Similarly, some religious and political beliefs can convert almost any seemingly irrational act into a rational one. The fact that some beliefs can convert otherwise irrational acts into rational ones, however, does nothing to show that all irrational acts must be based on false beliefs. A depressed person may act so as to harm himself, like Mr. K in Case 2-2, knowing meanwhile that he is acting irrationally. He may even state that he is unable to control himself. We all regard such an action as irrational even before we decide whether or not there are any false beliefs responsible for it.

Rationality as maximizing the satisfaction of one's desires

The inadequacy of the intellectualist account of rationality, which ignores irrational desires, has led to another account of rationality: that what is essential to rational action is the maximization of a person's desires. On this account an action is rational if it, better than any alternative course of action, maximizes the satisfaction of the person's desires. This position claims to account for the irrationality of a person who knowingly acts to harm himself: the person is acting contrary to what he knows will maximize the satisfaction of his desires, and hence his action counts as irrational even though he may have no false beliefs. This position does not ignore the importance of false beliefs, but regards actions based on false beliefs as irrational primarily because such actions do not normally lead to the maximization of desires for the actor.

Almost all contemporary moral and political philosophers, economists, and social and political scientists hold some version of this account of rationality as the maximizing of one's desires. (For example, see Rawls, 1971, pp. 142–150 and 395–452.) It accounts for almost all of the actions that we call irrational. Also, its proponents consider it advantageous that this view can be stated in purely formal terms, since it does not say anything about the content of one's desires. It requires only that one act so as to maximize the satisfaction of one's desires, regardless of their content, and thus seems to avoid arbitrariness and dogmatism. The most adequate version of this position is what we call the "cool moment" view: what is rational is maximizing the satisfaction of those desires that one has in a cool moment, taking into account the intensity of one's desires, their probability of being successfully satisfied, their effect on one another, and so on. One implication of this view is that an action which is rational for one person may be irrational for someone else, because their desires differ. This is as it should be. It would be irrational for a person whose most important desires involve his family to take a job which requires that he almost never be with them; while for a single person whose most important desire is career advancement, taking the same job might be perfectly rational.

But an adequate definition must not merely fit almost all the cases, it must fit every one, at least every clear one. It is not an adequate definition if one clear counterexample can be given. Thus we cannot seriously define a human being as a featherless biped because a plucked chicken could then be counted as human. Similarly, if we can find an example of an irrational action, one that everyone would regard as clearly irrational, that does not involve failing to maximize the satisfaction of one's desires, then we will have shown this definition to be inadequate. There are such cases. Consider someone with an overwhelming desire to mutilate himself in a very painful way. We may not know why the person has this desire—perhaps he is depressed—but we do know that he has no beliefs that such an action will come close to providing a compensating benefit for him or anyone else. Suppose further that as this desire gets stronger, his other desires get weaker, so that the time comes when, even taking into account all of his other desires, that action which would come closest to maximizing the satisfaction of his desires is to mutilate himself in a very painful way.

According to the maximization of desires view, such an action would now have to be called rational, but there seems no question that it is instead a paradigm case of an irrational action. One may object that anyone who would hold such a desire, especially with such intensity that its satisfaction comes closer to maximizing the satisfaction of his desires than any alternative, must be mentally ill. But this objection shows the inadequacy of defining rational action in terms of the maximization of desires, for the definition of rationality must account for the actions of the mentally ill as well as for those of everyone else.

In their account of rationality, philosophers have generally ignored those who are mentally ill; this is what has allowed them to think that such simple formal accounts of rationality as the one under discussion were adequate. The realization that some people have desires such that maximizing their satisfaction would result in their suffering significant harm, with no compensating benefit to anyone, makes the maximization of desires view much less plausible.

It is important to note that "mentally ill" and "irrational" are not only not synonyms, they do not primarily apply to the same sorts of things. It is persons who are mentally ill; it is actions that are irrational. Further, someone can be mentally ill but have no irrational beliefs or desires and perform no irrational actions. It is true that mental illness sometimes manifests itself in irrational actions, beliefs, or desires, but their irrationality is determined by independent criteria.

A defender of the position that acting rationally is maximizing the satisfaction of desires might claim that someone who is mentally ill cannot be in a cool moment when he considers his desire to mutilate himself as the most important of his desires. He may claim that the mere fact this desire seems most important counts conclusively against his being in a cool moment. But if one claims this, then the cool moment view has lost what its supporters regard as a crucial feature: it is no longer free of content. Rather, a cool moment has been defined, at least partially, in terms of the absence of certain specific desires, for example, the desire to mutilate oneself. Thus defined, the cool moment view is no longer a maximum satisfaction of desires position, for certain desires are ruled out from the very start as irrational. Upon closer inspection of the maximum satisfaction of desires position, it becomes clear that its power and plausibility all along depended upon ignoring certain desires. It is a plausible position as long as the desires of the person fall within the range we call normal, but given a desire like the one we are discussing, the view cannot handle it.

Rational action, though it may involve aspects of the intellectualist view and the maximum satisfaction of desires view, is not adequately defined by either. What is needed is at least some mention of specific irrational desires, whose satisfaction is irrational although no false beliefs are involved or even if their satisfaction will maximize the satisfaction of the agent's desires.

Defining irrationality

We disagree with the above positions in two important respects. First, we regard irrationality as a more basic concept than rationality; second, we hold that the definition of irrationality must have a specific content. We will give definitions of irrational actions, beliefs, and desires; those that are *not* irrational will be called rational. When we are not sure that an action (or belief or desire) is irrational (and there will inevitably be such borderline cases), we will call it rational. Irrationality is an important concept in the remainder of the book, and we want our definition to label actions, beliefs, and desires as irrational only when there is likely to be nearly universal agreement that they are indeed irrational.

Irrational action

We will take irrational action as our central concept and discuss irrational desires and irrational beliefs subsequently. To begin, a simple definition is as follows: *Irrational action consists of harming oneself without an ade-*

quate reason. In order to ensure the appropriate content, we must specify what is meant by "harming oneself" and by "adequate reason."

Harming oneself. Harming oneself means causing (or not avoiding) some evil for oneself. Evils or harms consist of the following conditions: death[1]; pain (physical or mental); disability (physical, cognitive, or volitional); and the loss of freedom, opportunity, or pleasure. We will discuss the concept of an evil in much greater detail in Chapter 4; here it is sufficient merely to list the evils.

To repeat, one harms oneself when one inflicts an evil on oneself. Sometimes one inflicts an evil directly, as when a person intentionally kills himself or causes himself physical pain, for example, by gouging his skin with a knife. But often a person acts in a way which he knows will result in his suffering an evil, though that may not seem to be his intention, as by frequently and compulsively washing his hands though it is obvious that continuous skin damage and pain is occurring. More normal persons also harm themselves by acting in ways which they know will result in their experiencing disabilities. For example, smokers with severe emphysema often continue to smoke while knowing that this will result in increasing physical disability and a greatly increased probability of early death.

We are not claiming that all actions in which one harms oneself intentionally or knowingly are irrational actions. Only if a person harms himself without an adequate reason is he acting irrationally. Further, unless self-harm is either directly or indirectly involved, the action is not irrational, regardless of whether one has any reason at all for doing it, let alone an adequate reason. It was taking "rational" rather than "irrational" as the basic term which led to the mistaken view that in order to be acting rationally, one must always have a reason for one's action. If I impulsively pick a flower from my garden, then unless this results in some foreseeable harm, my action is rationally allowed even though I had no reason for picking the flower, but simply felt like doing it. It is only actions which increase the probability of self-harm that need reasons in order to be rational. To cut off one's hand impulsively simply because one feels offended by it is irrational; in order for it to be rational, one needs some reason.

Some may object that it is not only harming oneself that needs a reason; they may claim that harming others without a reason is also irrational. There is no doubt that those who kill or harm others without a reason are often acting irrationally. The question, however, is whether harming others, like harming onself, is in itself an irrational act, or whether it is irrational because it is related to harming onself. Of course, harming those you love is irrational, for that results in your own suffering, but what about harming strangers? Though we talk of "senseless killing," it is not clear we regard the killers as acting irrationally unless we not only fail to see how they benefit from their act but also see that they have increased their own chances of being harmed. When those who harm others have no chance of being harmed themselves, we are outraged by their monstrous behavior, which we regard as immoral rather than irrational. For example, Nazi concentration camp guards who often indulged their cruel impulses to hurt their victims are not regarded as acting irrationally, but inhumanly or barbarously.

Reasons. For us, a reason must be a belief that can make an otherwise irrational action, that is, an act of self-harm, rational. The only beliefs that can do this are beliefs about the future—beliefs about the effects, either direct or indirect, of the action under consideration. Reasons always look to the future, and that is why they are often contrasted to emotions, which lead us to act on the basis of the past. "Feeling like doing something" does not count as a reason for doing it. We think that those who have regarded it as a reason have done so because they mistakenly believed that in order to act rationally, one must always have a reason for one's action. Freeing ourselves from this mistake allows us to regard most actions done simply because one feels like doing them as rational, even though "feeling like doing something" is not a reason. The problem with taking "feeling like doing something" as a reason becomes clear in those cases in which one does need a reason in order to act rationally. Suppose someone mutilates himself simply because he feels like doing so. Such an action is completely irrational: he has no reason for his act of self-harm.

It is important not to confuse irrational behavior with unintelligible behavior. People are often able to understand why someone acted irrationally. Sometimes our emotions are so strong that they lead us to act irrationally. Suppose a young father who has just lost his wife is so overcome with despair that he tries to kill himself. Though we may find such behavior intelligible or understandable, this does nothing to make us regard it as rational. A more trivial example is the golfer who, having hit a number of bad shots, becomes so angry that he throws his new set of golf clubs in the lake. These examples make it clear that present emotions do not count as reasons for actions, nor do facts about the past.

But what kinds of beliefs about the future count as reasons? First of all, the belief need not be true; we can be mistaken about the future and often are, but so long as our belief is not irrational, it may still count as a reason. (Irrational beliefs are discussed later in this chapter.) Whether or not it is a reason depends upon the content of the belief. We maintain that only beliefs that one's action will help oneself or someone else to avoid or relieve some evil or to gain some good, that is, to gain more ability, freedom, opportunity, or pleasure, count as reasons. One need not believe that one's action will directly result in the avoidance of evil or the promotion of good. It is sufficient if one believes that one's action is in accordance with some rule or practice which has the effect of lessening evil or increasing good, for example, that one promised to do the action.

Briefly, nothing counts as a reason for acting except a belief that someone, not necessarily oneself, will benefit, either directly or indirectly, by that action. This does not mean that any action which is *not* based on such a belief is irrational. As we noted before, in contrast to almost all previous accounts of rationality, we do not hold that one needs to have a reason to act rationally. One needs to have a reason (to act rationally) only when one is doing something to harm oneself. In all other cases, whatever one is doing can count as rational, and one needs no reason whatsoever for doing it. However, one can have a reason for an action that does not need one. For example, I need no reason for taking a walk; I may simply want to take a walk, and there is nothing irrational about my doing so. Yet I may also have a reason for taking a walk; I may believe that it is good for my health. Thus, though a reason is defined as a belief that can turn an otherwise irrational act into a rational one, it does not follow that this same reason cannot support acting in ways that are perfectly rational even without a reason.

Adequate reasons. We must now make clear what counts as an adequate reason. Since a belief is a reason if it can make an otherwise irrational action rational, a belief can be identified as a reason independently of context. Indeed we can actually give a list of reasons; for example, beliefs that I will earn some money, that someone else will be spared some anxiety, that I will increase my life span, and that I will make my children happy all count as reasons. Whenever I have any of these beliefs about any action of mine, I have a reason for doing that action, regardless of whether or not that action would be irrational without a reason. But it is only with regard to actions that involve harming oneself that one can speak of adequate reasons. (When discussing morality, actions that involve harming others need morally adequate reasons.) An adequate reason depends upon the context. It is a reason that is adequate to make some particular self-harming action rational. It should be obvious that such a reason need not be adequate to make some other act of harming oneself rational. That I will earn some money or make a child happy for several hours is, under normal circumstances, an adequate reason for undergoing mild pain for a brief period, but earning the same amount of money or causing the same amount of happiness is not, again under normal circumstances, an adequate reason for risking death or a serious disability.

If I cause my leg to be amputated in order to get rid of an annoying Plantar's wart, that would count as a very irrational action because I am inflicting on myself several very great evils: pain, permanent disability, and increased risk of death (through a wound infection) for an inadequate reason. However, if I allow my leg to be amputated because I have a malignant osteosarcoma in the femur and amputation may prevent metastatic spread, then I have an adequate reason for acting, because the pain, disability, and probable death associated with metastasized bone cancer are at least as significant as the evils associated with the amputation.

Similarly, a very depressed man may willingly take a tricyclic antidepressant drug, knowing very well that it may cause such evils as a dry mouth, mild visual blurring, and mild dizziness when arising from bed. If the patient believes that taking the drug will probably relieve the evil of the depression, then he has an adequate reason for taking it.

As mentioned above, adequate reasons may also include the relief of evils for others. For example, you may rationally suffer the evil of moderate personal deprivation associated with giving up a fourth of your income for the adequate reason that giving the money to your disabled penniless brother enables him to avoid the evils associated with great poverty.

Adequate reasons may also include beliefs about gaining some good for oneself or others. Consider a premedical student who dislikes biology but elects to major in the subject; her reason would be considered adequate if she believed that her chances for medical school admission were thereby significantly enhanced. She would be electing to suffer the unpleasantness associated with taking courses she disliked because of her belief that she could subsequently obtain the many goods associated with being a physician.

In order for a reason to be adequate, rational persons would have to rank the evils I or someone else will probably avoid or relieve through my act of self-harm (or the goods that will probably be obtained) to be as great as the evils that I cause for myself by this act. Judging the adequacy of a reason always involves a balancing of evils versus evils (or evils versus goods) of this kind.

In the case of the depressed man, the balancing involves comparing the possible evils of a dry mouth, visual blurring, and dizziness caused by taking the drug with the evils of a possibly prolonged severe depression if he does not. A prolonged severe depression would seem to everyone a much greater evil than the minor annoyances usually associated with tricyclic drugs. Therefore, the probable relief of the depression seems an adequate reason for exposing oneself to the side effects of the drug, and therefore taking the drug does not seem irrational. In fact, failure to take it seems irrational, unless there is some further reason not yet mentioned, such as an alternative treatment that is equally effective.

It is very often true that the evils one hopes to avoid or ameliorate (or the goods one hopes to attain) are only probable while the evils one inflicts on oneself for this purpose are often certain, for example, when one consents to an amputation in the hope of arresting a cancer. However, sometimes both the harms caused and the harms ameliorated are less than certain. The depressed man taking the tricyclic is likely (70–80 percent) to experience at least some side effects, but it is only probable (60–70 percent) that the drugs will relieve his depression. Since the two sets of evils differ so much in intensity and the probabilities are fairly close, the probabilities here do not significantly affect the balancing, and his taking the drug is not irrational. This case illustrates another very important feature of the balancing of evils: it is not just the *intensity* of various evils that enters into balancing but also their *probability* of occurrence. Later in the book, in Chapter 9, for example, we will see many cases in which the vast differences in the probability of occurrence of various evils play a crucial role in judgments about the irrationality of a person's actions, and about the consequent justifiability of various psychiatric interventions.

Rationally required actions versus rationally allowed actions. An adequate reason makes an otherwise irrational action rational. It does not necessarily make abstaining from that action irrational. For example, let us return to Miss A in Case 2-1. If she chooses not to kill herself, she will postpone the great evil of death by several more months, but she will experience the great evil of intense pain. If she does take the barbiturates, she will inflict the great evil of death on herself but will thereby avoid the great evil of several months of intense suffering. Thus both courses of action involve the self-infliction or the nonavoidance of some great evil, but in each case there is an adequate reason for doing so (either the avoidance of intense suffering or of death), so neither course of action seems irrational. We would say that either course of action is rationally allowed. Which one Miss A chooses will depend on which she regards as the lesser of two evils.

On the other hand, if I can probably save my life by having my arm amputated and if I will certainly die without the surgery, then given normal life prospects, having the surgery seems rational (the attendant pain and disability being compensated for by the probable avoidance of death), while not having the surgery seems irrational (since I would be choosing, without an adequate reason, not to avoid death). Therefore, if these are the only two options open to me, it follows that having the surgery is rationally required. (It is important to note here that just because the surgery is rationally required, this in and of itself does not justify others in forcing the surgery on me without my consent. The requirements which must be met before patients can justifiably be treated without their consent are very stringent and are not met *simply* by the fact that choosing treatment would be rational and refusing treatment would be irrational. This topic is discussed in Chapter 8.)

An adequate reason can sometimes be so strong that it requires my acting in accord with it; if I do not do so, I am acting irrationally. However, adequate reasons are not generally this strong. In order to be adequate, they must be strong enough to make the particular act of self-harm rational, but they can easily be this strong without being so strong that they make a refusal to act irrational. Indeed, some reasons, those involving the welfare of others, though they may make many acts of self-harm rational, never make a refusal to harm oneself irrational. It is always rationally allowed, though it may not be morally allowed, to prefer my own benefit to that of others even when a small sacrifice on my part may lead to the avoidance of a great sacrifice by many others. Of course, it is also rationally allowed to make some self-sacrifice in order to benefit others and this is generally regarded as morally praiseworthy. Thus though there is no ambiguity in the term "irrational," the term "rational" contains two important subcategories. What is rational may be rationally allowed, which means that it is rational either to do or to abstain from doing the action. What is rational may also be rationally required, which means not only that it is rational to do the action, but that it is also irrational to abstain from it. The vast majority of acts that we perform, though rational, are not rationally required; they are merely rationally allowed. Recognition of this fact should make people less inclined to interfere with the choices made by others.

Rationally allowed treatment choices. Frequently in medical practice, *either* consenting or refusing a particular suggested treatment would be rational. Also, a patient is often faced with two (or more) treatment alternatives, either of which would be rational.

Consider first those cases in which either consenting or not consenting to a particular treatment would be rationally allowed. Often these are cases in which the evil being suffered is relatively mild and the treatment itself is accompanied by only mild evils. For example, suppose I have a small and only mildly painful Plantar wart on my foot. I have had such a wart before, it never became more painful than it is now, and eventually it went away on its own, though only after six to eight months. I could choose to have the wart removed immediately by surgery, which would cause me a moderate amount of pain and inconvenience for one to two days, but there would be a very high probability of no wart at all after that (that is, no regrowth within the immediate future). In cases like these, it is rational either to choose or to decline treatment, and in fact such procedures are commonly called "elective" by physicians. Of course, if my wart grew and became very painful and disabling, at some point it would become irrational not to have this very safe and only mildly painful minor surgery. In many cases of cosmetic surgery as well as in a host of minor surgical procedures, it is rational either to have or not to have the operation.

Of potentially greater moral significance are those cases in which patients are faced with two (or more) treatment options, either of which involves the probable suffering of significant evil and either of which would be a rational choice. For example, some patients with back problems have the option of medical versus surgical treatment. The choice will depend on a complex balancing in each case of such factors as length of treatment process, probable relief of symptoms (evils), probability of recurrence, morbidity associated with the treatment process, and so on. In many such cases, either treatment would represent a rational choice, and the selection should be left up to the patient's own ranking of the probable evils involved in the available options.[2] Many cases of seasonal allergies also fall in this category. Here, the patient's own ranking of evils ought to determine whether symptomatic treatment (antihistamines, etc.) should be used every year or whether desensitization treatment should be attempted. Freud talked about some neuroses that might better be left untreated, for the pain of treatment could be greater than the suffering caused by the neurotic symptom. We believe that there are many cases in which it may be rational for a person to refuse psychiatric help for minor suffering because of the associated monetary and emotional costs.

Sometimes a physician will be aware that two or more rational treatment options are available but will inform the patient of only one. On rare occasions, the physician's motives may be purely self-serving, for example, the surgeon who, because he wishes to operate, doesn't inform his patient of a nonsurgical treatment which he knows others regard as equally efficacious. But much more common is the instance in which the physician does believe that the treatment he recommends would be much more beneficial for the patient than another treatment which is not even mentioned, though he is aware that his belief is not strongly supported and that others, equally well informed, favor an alternative treatment. If it would not be irrational for the patient to choose this alternative treatment (if he knew about it), then the physician is, in almost all cases, morally obligated to inform the patient about it. We will show in Chapter 8 why not informing in cases like these usually represents unjustified paternalism.

We have mentioned the cases of back problem and seasonal allergy as often involving two rational treatment options. Let us work through a third example, of a woman with breast cancer, to show how a physician may, with sincerely benevolent motives, deprive a patient of the opportunity to choose a completely rational treatment option. Suppose the woman could have either of two operations. One would be a more radical and disfiguring operation which would also probably leave her with at least a mild degree of postsurgical disability (such as impaired arm raising) and an estimated five-year survival rate of 65 percent. The second operation would be much less radical and disfiguring, and would also probably result in less disability, but would have an estimated five-year survival rate of only 60 percent. Her physician places a high personal value on survival itself, assumes the patient does too, and mentions only the more radical operation. It is important to see that it would not be irrational for the woman to elect the less disfiguring operation. The two treatment options contain different kinds and possibilities of evils. This particular woman may place a high personal value on her physical integrity and appearance and, in order to preserve it, be willing to accept a somewhat lower probability of five-year survival. To choose the evil of a 5 percent chance of not surviving five years in order to avoid the evils of greater physical disfigurement and mild disability does not seem irrational. Thus, in this kind of case, the choice of type of surgery should properly be the patient's.

Irrational desires

So far, we have concentrated on providing an account of irrational actions. However, it is not merely actions that can be classified as irrational; some desires are irrational. Though irrational desires are very closely related to irrational actions, it is useful to give an explicit account of them. We have already noted that an irrational action is one in which a person harms himself without having an adequate reason for so doing. One simple way of characterizing an irrational desire is to say it is a desire to carry out an irrational action. This means that an irrational desire involves both wanting to suffer some evil and *not* having an adequate reason for doing so. Thus we cannot know whether the desire to have one's arm cut off is irrational unless we know whether or not the person has an adequate reason for it. If he does have such a reason, for example, preventing the spread of cancer and thus preventing his death, then his desire for amputation is not irrational. But if he has no reason but simply an overwhelming desire to do so, or voices have told him to do so, or he believes that he deserves to have his arm cut off, then his desire is irrational. And if he has an inadequate reason for wanting to have his arm cut off, for example, winning a bet of ten dollars, or he will avoid the unpleasant necessity of participating in physical education courses, then his desire is also irrational. Briefly, if doing something is irrational, then having the desire to do it is also irrational.

It is important not to identify irrational desires with the desires that are involved in compulsions and phobias. One can have irrational desires that are not part of any compulsion or phobia, and the desires that accompany phobias and compulsions need not be irrational. Of course, a phobia or compulsion often does involve irrational desires. One may desire to do something, for example, wash one's hands repeatedly, knowing that doing so will involve injuring one's hands, without having an adequate reason for wanting to do this. This is because one has no conscious reason for wanting to wash one's hands; one simply has a very strong desire to do so.

However, it is possible for someone who has a hand-washing compulsion to resist the compulsion for a period of time and thereby to suffer extreme anxiety. Subsequently, he may wash his hands with the belief that he will avoid the extreme anxiety that results from not doing so. If he consciously washes his hands in order to avoid the anxiety that results from not doing so, then his desire is not irrational and neither is his action. That does not mean that he has no illness or malady. It simply means that given his condition, he is acting rationally. He still is suffering from a compulsion, and this is a mental malady, but his actions are no more irrational than that of the person who loves chocolate ice cream yet refuses to eat it because it causes an allergic rash. This person is depriving himself of pleasure, but his action is rational because his allergy provides him with an adequate reason for depriving himself of pleasure. In the same way, the person with a compulsion is acting rationally in causing himself some pain and injury if he is doing so to avoid intense anxiety, because this provides him with an adequate reason for harming himself.

But if a person is unaware that he will suffer intense anxiety if he refrains from washing his hands, and so has no conscious belief that he is in any way benefiting from repeatedly washing his hands, then, given that he knows he is harming himself, he is acting irrationally. Whenever I have no conscious belief that anyone will benefit from my action, I have no reason for my action. If I have a conscious belief that someone will be harmed by my action, that may provide a motive for my action, but it is not a reason. We distinguish sharply between reasons and motives; in order to count as a reason, a belief must be capable of making some otherwise irrational action rational. In order to count as a motive, a belief must merely count as part of the explanation for one's performing the action.

Suppose, for example, that someone has harmed a member of my family and I want revenge. Suppose further that the only way in which I can harm that person is to inflict the same harm on myself, that is, I can kill him only by also killing myself. Killing myself is an irrational action unless I have an adequate reason for doing so; in the present case, however, not only do I not have an adequate reason, I have no reason at all. My motive for killing myself is my belief that I will thereby kill him, thus enacting revenge, but I have no beliefs that anyone, myself included, will benefit either directly or indirectly from this act. Hence, though I have a motive, revenge, I have no reason for my action, and it remains irrational.

If we consider only conscious beliefs, then most motives are reasons; it is primarily motives which involve harming onself or others that do not count as reasons. When we consider unconscious motives, it becomes very important to note that they do not count as reasons.[3] It is not merely that reasons are conscious beliefs and that unconscious motives are unconscious beliefs. It is more to the point that unconscious motives usually are not beliefs that someone will benefit from one's action, but rather the reverse, that that person will be harmed by it. Unconscious motives are beliefs that we cannot consciously acknowledge because they conflict with our conscious goals, values, ideals, and so on. It should not be surprising that beliefs about one's actions harming oneself or others generally motivate our actions unconsciously rather than consciously.

Motives, especially unconscious motives, are often attributed to a person to explain an irrational act, especially when the person appears to have no adequate reason for the action. For example, consider a very tense and unhappy man who from time to time mutilates his body by gouging his forearms and calves with pieces of broken glass. When questioned about his action, he says only that he is sometimes overcome by the urge to cut himself. This would clearly count as an irrational action since he is harming himself without an adequate reason; that is, he does not appear to have any conscious belief that his self-harm will avoid any evil or gain any good for himself or others. We are then apt to try to explain his irrational behavior by attributing an unconscious motive to him (say, a belief that he deserves to be punished), especially if this motive is consistent with other facts we know about him. But though the motive does have explanatory value, it is not a *reason* according to our definition of a reason, since it is not conscious and it is not a belief about avoiding an evil or gaining a good. Suggesting that someone believes he deserves to be punished may make his self-mutilation *understandable*, but it does not make it *rational*.

Irrational beliefs

We have noted that the definition of both irrational actions and irrational desires involves the notion of "adequate reasons." Earlier, we gave an account of reasons and of adequate reasons and pointed out that reasons must be conscious beliefs. In that section, we did no more than note that these beliefs must also be rational. The belief that I will prevent my child from dying is a reason for mutilating myself only if it is rational. Thus in order to complete our account of irrational actions, it is necessary to discuss irrational beliefs. Though irrational beliefs can serve as motives for actions in which the person harms himself, they can never be reasons, and these actions must therefore always be regarded as irrational.

What is an irrational belief? As we use the phrase, it is closely related to the psychiatric term "delusion." However, delusions are generally regarded as symptoms of some psychotic condition, whereas we use the phrase "irrational belief" such that it is irrelevant whether or not the belief is related to a psychosis. Thus we consider delusions a subclass of irrational beliefs, those irrational beliefs that are manifestations of some psychotic condition. Obviously the analysis of irrational belief will be closely related to the analysis of delusion.

First, it is necessary to specify that the notion of an irrational belief is person dependent; what is an irrational belief for one person need not be one for another—it may merely be a mistaken belief. It is not irrational for a four-year-old to believe in Santa Claus, but that belief would be irrational for any adult of average intelligence and experience. We can specify this person dependence by the following formulation. A belief is irrational only if it is held by a person with sufficient knowledge and intelligence to know that it is false.

Does this mean that an irrational belief must be false? This is the standard view. Indeed, delusions are sometimes simply defined as false deviant beliefs. As Moor and Tucker (1979) have recently pointed out, however, it is not necessary that the belief is false but only that the belief is contrary to the overwhelming evidence available to the person and that he has sufficient knowledge and intelligence to evaluate this evidence. No doubt 99.99 percent of beliefs which are contrary to the overwhelming evidence available to us are false, so it is not surprising that one is tempted to regard irrational beliefs as necessarily false. But an irrational belief held by a paranoid schizophrenic does not cease to be irrational if, quite fortuitously and completely unknown to him, someone *is* plotting against him. It is the fact that the belief is held in the face of overwhelming evidence to the contrary that makes it irrational. And it is this feature that we regard as central to irrational beliefs.[4]

The third feature of irrational beliefs merely clarifies the relationship between the first two features. It explains why irrational beliefs are regarded as deviant beliefs. We formulate this third feature as follows. Irrational beliefs are those whose contradiction by the overwhelming evidence is obvious to almost everyone with knowledge and intelligence similar to that of the person holding the irrational belief. This third feature ensures that we do not use the phrase "irrational belief" merely as a way of disparaging beliefs that we, with our superior knowledge, find obviously false. For example, it means that we cannot call religious beliefs irrational if they are in fact held by a number of people. This is as it should be, for to call a belief irrational is to say something about the mental state of the person holding the belief; it is a strong indication of a mental or emotional problem. But for someone to hold a religious belief shared by a number of others indicates nothing psychiatrically significant about a person's mental or emotional state.

Another important feature of this account of irrational beliefs is that it allows for a belief to become irrational as evidence accumulates against it. Thus a mother may believe that her son who has been reported missing in action is still alive, and this belief need not be regarded as at all irrational. When she receives a notice that he has been found dead, her continued belief that he is alive may be regarded by many as irrational, though given the fact that mistakes are not rare, this could be thought borderline. However, if ten years pass with no contrary evidence, her continued belief that he is still alive seems clearly irrational. This example also brings out an important feature of our account of irrational beliefs: there are borderline cases and disagreements about whether a given belief counts as irrational, though the overwhelming number of cases will be clearly rational or irrational. Thus our account seems to agree with the standard practice of classifying psychotic delusions and severe cases of denial as irrational beliefs.

One important question that needs to be addressed is why we use the label "irrational" for a belief with the three features mentioned: (1) it is held by someone with sufficient knowledge and intelligence to know that it is false, (2) it is contrary to the overwhelming evidence available to that person, and (3) its contradiction by the overwhelming evidence is obvious to almost everyone with similar knowledge and intelligence. One can say that this is just what is meant by an irrational belief, and in large part we agree. But both irrational actions and irrational desires have an intimate connection to one's suffering some evil with no compensating benefit to anyone, and irrational beliefs, as we have defined them, seem to have no connection with suffering any evil at all.

It should be clear on closer inspection that irrational beliefs are closely connected with an increased risk of suffering evils. Actions based on such beliefs have a high probability of resulting in suffering unnecessarily; that is, such beliefs often result in irrational actions. For example, if I irrationally believe that I am intensely wicked and must suffer for my sins, I am apt to engage in actions that will themselves cause me to suffer evils: to stop seeing friends because I believe I do not deserve their time; to pass up a new employment opportunity because I believe I do not deserve success; or even to kill or injure myself. Similarly, if I hold the irrational grandiose belief that I am invincible and bound to succeed in whatever I attempt, I am apt to behave in ways toward others or to embark upon financial or other projects that will eventually cause me great loss and suffering.

Another way in which irrational beliefs are closely tied to self-harm is that the very holding of such beliefs is frequently associated with anguish and suffering. This is certainly true with irrational beliefs about self-worthlessness. It is also true of many irrational beliefs that may be present in schizophrenia, such as beliefs that one's thoughts are controlled by or audible to others, or that one's body is rotting. However, irrationally grandiose beliefs about oneself, at least when they are mild to moderate in degree, are apt to be associated with more euphoria than anguish; it is the indirect, secondary effects of holding and acting on grandiose beliefs that results in self-harm.

Notes

1. We use the term "death" to refer not only to biological death but also to loss of personhood, that is, complete and irreversible loss of all consciousness and cognition. (See Chapter 10 for further discussion of these concepts.) Thus we regard death not as a positive evil like pain, but rather like the loss of pleasure, or freedom; but it is the loss of everything.
2. See Weinstein, Fineberg, Elstein, Frazier, Neuhauser, Neutra and McNeil, 1980, for a sophisticated statistical treatment of decision making in the clinical situation.
3. See Gert, 1967, for a fuller discussion of unconscious motives.
4. See Chapter 6 where holding an irrational belief is related to volitional disabilities.

References

Gert, Bernard. Hobbes, mechanism and egoism. *Philosophical Quarterly*, 1967, *15*, 341–349.

Moor, James H., and Tucker, Gary. Delusions: Analysis and criteria. *Comprehensive Psychiatry*, 1979, *20*, 388–393.

Rawls, John. *A Theory of Justice.* Cambridge: Harvard Univ. Press, 1971.

Weinstein, Milton C., Fineberg, Harvey V., Elstein, Arthur S., Frazier, Howard S., Neuhauser, Duncan, Neutra, Raymond R., and McNeil, Barbara J. *Clinical Decision Analysis.* Philadelphia: Saunders, 1980.

3

Valid Consent and Competence

This chapter will discuss the features of valid consent, more commonly called "informed consent." We prefer the phrase "valid consent" because it suggests a parallel with the making of a valid will, a parallel we shall explore further in this chapter, and because the phrase "informed consent" is misleading as it suggests that all that is necessary for valid consent is adequate information. We think that valid consent is a more complex notion than has generally been recognized. It involves, in addition to the notion of information, the important concepts of coercion and competence. Further, we believe the concepts of coercion and adequate information, as well as that of competence, require more analysis than they have typically received.

Before we discuss the elements of valid consent, it is important to discover what kinds of situations require it. Though it is in medical practice that the notion of valid consent is now most prominent, the concept is not limited to medicine. In any situation in which someone contracts for services with a professional with whom he has a fiduciary relationship, the notion of valid consent may come into play. If a lawyer or an accountant suggests a particular course of action to a client, and the client authorizes this action on his behalf, then in order for the authorization to be valid, the same features must be present as are necessary for valid consent in medicine. It is important to make this point, since if one believes that valid consent is required only in medicine, one may not only distort the concept, one may also foster the incorrect view that ethical considerations are different for doctors.

There are, of course, reasons why the concept of valid consent has come to play a larger role in medicine than in any of the other professions, but these are matters of degree, not of kind. First, medical patients often know less about the alternatives to, and consequences of, the courses of treatment suggested to them than do the clients in other professions. Second, because of their condition it is often more difficult for them to understand the information provided. Third, the consequences of these courses of treatment are often more serious than are the consequences of actions involving other professions. Fourth, the medical profession has a history of making decisions for patients without their valid consent to a much greater degree than the other professions. We can summarize these points as follows: in medicine, it is often more difficult to achieve valid consent because of the amount of information needed and the condition of the patient; it is often more important to obtain it because of the serious consequences to the patient; and often little attempt has been made to obtain it because of the paternalistic attitude of the medical profession.

Adequate information

Ideally, a patient should know everything that would affect his decision concerning which of the courses of treatment available to him he should choose. This means, first, that he should be informed of everything that all rational persons would want to know: all of the goods and evils involved in the various alternatives, including their severity and probability. In addition, each patient should know of anything else that might affect his *personal* decision. He may have religious beliefs, or cultural beliefs, or even superstitions that would influence which alternative he would choose. Ideally, he should know of any facts that would lead him to decide one way or another, even on the basis of these superstitions. What he does not need to know, unless he happens to have some special interest in the matter, are technical details concerning the treatment.

Thus the adequate information component of valid consent does not demand, even ideally, that the patient know the chemical formulae of his medications, the internal location of any incisions, or even the location of the various affected internal organs, for example, the spleen or the liver. Though these are appropriate objects of patients' curiosity, and providing such information often relieves a patient's anxiety, they normally have no relationship to valid consent. Adequate information must include only those facts that all rational persons would want to know, namely, the various goods and evils that result from the alternative modes of treatment, including their severity and probability. However, in some cases, it must also include those facts that would affect the decisions of persons with idiosyncratic views; for example, if a patient has views that prohibit being seen nude by members of the opposite sex, then he must be told if a procedure requires being thus seen.

One may object that medicine is an inexact field, that we do not know with any precision the goods and evils which might result from alternative treatments, that we do not even know the probabilities except in the vaguest fashion. But this objection has little force, for one who puts it forward is subject to the following dilemma. If knowledge is as inexact as you say, then how can you rationally decide which of the various alternative treatments to recommend? Whatever estimates you make of the risks and benefits involved, including the knowledge that the probability estimates are not reliable, can be conveyed to the patient. Unless one wishes to defend purely intuitive judgments that cannot be rationally justified even to colleagues, then it must be possible to convey the grounds of one's judgment to the patient. Further, making the risks and benefits of the various alternatives explicit helps counter the various biases that enter into making one's judgments. To guard further against the effects of these biases, patients must be told if there is legitimate disagreement among medical practitioners about the value of the different treatment alternatives.

Except in unusual circumstances, it is not difficult to determine what should be told to a patient: namely, everything that one has any reason to believe would affect his decision. It is the patient's life, health, time, money, and so on that are most directly affected by his decision, and thus he is entitled to make the decision on the basis of the most complete and personally relevant information possible.

But even though the kind of information a patient must have is limited to information about the goods and evils, their severity and probability, and to information relevant to a patient's special values, it is still impossible to give complete information. There are an indefinite number of very small risks, say of the magnitude of one in ten million, about which it does not seem morally required to inform the patient. This has been recognized by almost everyone writing on the subject of consent. However, most justify not providing this information on the somewhat questionable grounds that the patient probably already knows, for example, that there is some extremely small but finite risk in taking diagnostic blood samples. We justify not telling about very small risks on the grounds that rational persons would not find them relevant to their decision. This justification also explains why it is not just the probability that determines if the patient should be told, but also the severity of the evil if it should occur. Thus, one need not tell about a one in ten thousand risk of a severe but temporary pain, but should tell if there were a one in ten thousand risk of death.

It is very important that a patient be informed of the risks and benefits of all aspects of a proposed course of treatment. For example, a patient confronting possible surgery should be given adequate information with regard to not only surgery but also anesthesia, if there are any significant differences in risks and benefits among the possible alternative methods. Pregnant women are often told of the various possible forms of anesthesia and allowed to choose among them, but other surgical patients are frequently not presented any such alternatives. This is often true for parents of children who are confronting surgery. These children can be sedated preparatory to anesthesia by giving them either an injection or an oral medication. There is no question that almost all children prefer a pill to an injection, though some anesthesiologists prefer the injection. However, except in rare cases in which oral medication is contraindicated, parents should be informed of the alternatives and allowed to make the choice. The anesthesiologist may certainly make recommendations, but he should not make the decision unilaterally without even informing the parents of the alternatives.

Another instance of information that is generally not given, but that is clearly relevant, is the mortality and morbidity rate for the contemplated type of procedure in one's specific hospital or medical center, especially if there is a significant difference between local results and those of medical centers where a particular kind of procedure is performed more often. There is now substantial evidence that some kinds of surgery are volume-sensitive to a rather high degree (Luft, Bunker, and Enthoven, 1979). Thus, all candidates for volume-sensitive operations at low-volume hospitals should be told that they are at a somewhat greater risk than they might be elsewhere. Any rational person who is advised to undergo a major operation, for example cardiac or major vascular surgery, would want to know that traveling fifty miles (or five blocks) might lessen his chances of dying during or after the surgery.

Of course, some patients might still elect to undergo volume-sensitive surgery at a low-volume hospital. They might be willing to take the risk because they prefer to receive treatment nearer home, because they like and trust the staff of a low-volume hospital, or for other reasons. Such a preference is not necessarily irrational. But the patient should be told of the relation between surgical volume and mortality before he makes his decision.

Some physicians may believe that their own low-volume hospitals (or practices) represent exceptions to the published volume-mortality data. For valid consent, one is still required to acquaint patients with the existence of such data and then to explain whatever local mitigating circumstances are believed to be present. Patients would then be free to decide for themselves where to have their surgery (see Culver and Gert, 1980).

One should also know the mortality and morbidity rate of one's surgeon, especially if it is significantly different from that of other surgeons in the same hospital, or in the same vicinity, who operate on cases of comparable severity and complexity. This assumes that such knowledge is available.

It is also relevant for the patient to know if the particular kind of surgery being suggested is performed far more often in the patient's area of the state or country than elsewhere, because this at least raises the possibility that the suggested surgery may be unnecessary, and that a second opinion from outside the area may be desirable. There is now strong evidence that the frequency with which particular kinds of surgery are performed (e.g., hysterectomy, tonsillectomy, and prostatectomy) varies widely from area to area (Wennberg, 1979; Wennberg, Bunker, and Barnes, 1980). Because individual physicians are often unaware of these kinds of epidemiological data, we would recommend that such data be collected and published by public health agencies and be made prominently available through local libraries, consumer's groups, and so on.

Some of the above-mentioned kinds of information go beyond what is now required by law. The fact that our account of valid consent requires a physician to tell patients of this information may seem evidence of the lack of practical impact of philosophy. However, one should note that what counts as information the patient should have before giving consent has usually been determined not by legislation but by court decision (see *Cobbs* v. *Grant*, 1972). A patient who was not given some of the information mentioned above and who suffered an unfortunate complication due to the surgery might decide to bring suit against his physician. If it were plausible that he might have made a different decision if he had obtained the information, and if it were established that the information was indeed available but was not given to the patient, there is a genuine possibility that the jury decision would be in his favor. It has already been determined by some courts that the information required for valid consent is determined not by the standard practice of physicians to tell, but by what reasonable persons would want to know (for example, *Cobbs* v. *Grant*, 1972). Thus, it may be only a matter of time before the law enforces what we believe is morally required.

It is clearly an historical accident that consent forms are required for surgery and not for most nonsurgical procedures. Side effects from drugs can result in complications as great as those from surgery. Clearly, these problems should be discussed as fully as the risks of surgery. There are also risks in nontreatment, and the patient should be informed if other qualified physicians would advise more aggressive treatment, whether surgical or medical. Some persons seem automatically to think that when both conservative treatment and surgery are possible the former is better, but that choice, just like the choice to refuse surgery, is up to the patient. Patients should be told of all of the legitimate alternative courses of treatment. All these issues simply make explicit the consequences of the view that the patient should make the decision, not the doctor.

An interesting example of a physician making a conservative decision without informing the patient of alternatives arises in psychiatry. For perfectly understandable reasons, electroconvulsive therapy (ECT) is treated like surgery, and a consent form is needed. Psychiatrists are often reluctant to use ECT to treat major depression until a course of antidepressant medication has been tried and failed. This conservative method of proceeding may have been in order before the recent developments in ECT which make it an extraordinarily safe and painless procedure. We are not recommending that patients routinely be given ECT for major depression before going through a trial with antidepressant medication; we are not recommending any particular type of treatment over another. We are recommending that patients be given the choice of ECT before embarking on a course of antidepressant medication. ECT has a success rate of roughly 80–90 percent with major depression, and antidepressant drugs have a success rate of roughly 60–70 percent. Further, it often takes at least ten days for the drugs to have beneficial effect, whereas ECT can have an effect in a shorter time. It is true that ECT produces a temporary loss of memory, though this is usually minimal with unilateral ECT (affecting only one hemisphere of the brain), and there is a very slight risk of some longer-lasting memory difficulties, though these have not been documented with certainty. (For a further discussion of the consent process in ECT, see Culver, Ferrell, and Green [1980].) Whether these risks are outweighed by the greater probability of success and the greater speed of action of ECT is something for each patient to decide for himself. The doctor should not preempt the decision by failing to present the patient with the two alternatives.

Of course, the psychiatrist can recommend drugs over ECT, or vice versa. We do not advocate that doctors remain silent and require patients to make their decisions unaided by expert medical advice. Depressed patients, for example, often have a difficult time making decisions, and they may simply wish to follow their doctor's advice. All that we advocate is that patients be made aware of the alternatives. No doubt this takes time, and as we all know, time is in short supply. If a physician is required to tell a patient of all the legitimate alternatives, he may have to spend more time with each patient than he now does. We do not question the practical difficulties in having physicians provide adequate information to their patients. It is possible that many physicians will not want to take the time to provide the kind of information required. Objections to providing adequate information to the patient, though often stated as impossible, are not impossible in any important sense; they simply demand a change in the system of providing information for valid consent. Quite possibly the provision of information should no longer be regarded as the sole responsibility of the physician. Instead, in many in-patient settings, it could become one of the major responsibilities of the nursing profession.

As noted earlier, the information that a patient should have is not technical; it is information about the severity and probability of the goods and evils of the recommended course of treatment compared to no treatment or other treatments. For almost all standard treatments, nurses could be trained to provide such information. Indeed, nurses are often as well suited as doctors to provide the information required for valid consent. They are possibly less likely to overwhelm the patient with irrelevant technical details and more likely to concentrate on that information which the patient really needs, for example, how much pain and risk of disability are involved, how long a hospital stay is anticipated, and how high the chances are of a successful outcome. Nurses could then determine that the patient did indeed have all of the relevant information required for valid consent. They are often in a better situation than doctors to determine how much the patient understands and thus to certify that the patient does have adequate information to give valid consent. Patients may more easily reveal their lack of understanding to nurses than to doctors. Making nurses in in-patient settings responsible for determining that the patient has adequate information for giving valid consent would help to institutionalize the nurse's role as patient advocate, and thus make it the nurse's professional duty to protect the patient from making decisions based upon inadequate information.

Refusal of information

We have discussed in some detail what information a patient needs and who should be responsible for determining that he has it. We now consider a very puzzling problem: what should be done when a patient refuses to listen to the information being provided, and, in fact, insists that he does not want to hear about the risks, alternatives, and so on, but simply wishes the doctor to decide what is best? We are not talking about the doctor or nurse *inferring* from the patient's behavior that he does not want to know about the risks and alternatives. This does not seem to pose a problem of the same complexity. We realize that doctors and nurses sometimes wish to spare their patients the anxiety of knowing about the risks of the procedures they must undergo and the difficulties associated with making decisions. However, in the absence of an explicit refusal to be told of the risks and alternatives, we think it is clear that the patient must be told.

A patient who explicitly refuses to listen to information that is relevant to his making an important personal decision, though unusual, is useful to consider because he allows us to state our views concerning valid consent in a very clear fashion. Competent patients should make their own medical decisions. However, if a patient is explicitly willing to waive his right to obtain adequate information before consenting to a particular medical procedure, then, provided there are sufficient safeguards, we believe such a waiver is acceptable. Of course, the doctor or nurse should make several attempts to provide the patient with the relevant information about the alternative modes of treatment, but if the patient steadfastly refuses to listen to this information and explicitly grants the doctor the power to decide for him, then we regard this explicit waiver of consent as an adequate substitute for valid consent. We make this point with some reluctance, knowing that the procedure could be abused as a way of short-circuiting the requirement for adequate information.

When there is only one medically acceptable treatment, it is clearly not irrational to refuse to listen to all the risks that this treatment involves. Hearing about these risks is unpleasant, anxiety-provoking, and so on, and since there is no real alternative to the treatment, in some cases not much is gained by knowing all the risks. On the other hand, it would also be rational to want to know what is likely to happen to oneself. For example, there may be nonmedical decisions that one would want to make if one knew that there were a significant risk of death or permanent mental disability. Thus, it is usually advisable for patients to accept sufficient information to provide valid consent even when there is only one medically acceptable treatment.

When there is more than one medically acceptable treatment, then waiver of consent may border on the irrational. Consider alternative treatments for cancer, such as surgery versus chemotherapy or radiation; here one may have the choice of different survival rates versus considerably different amounts of pain and/or disability. Or consider ECT versus trycyclics for depression, where speed and probability of relief are balanced against different kinds of side effects, or psychotherapy versus behavior therapy, where for some conditions there may be an extraordinary difference in probable length of treatment and very different claims about their benefits. To allow someone else to choose for you, unless you are very sure that they know your personal values and will choose in accordance with them, seems to expose you to the possibility of being subjected to what you would consider the greater of two evils, in order to avoid the unpleasantness and anxiety involved in making the decision yourself. But though it may border on the irrational, it is nonetheless not clearly irrational, and hence there seems to be no argument that would prohibit one from signing a waiver of consent and entrusting the power of medical decision making to one's doctor.

Coercion

Adequate information is not all that is required for valid consent. The consent must also be given without coercion. There is probably very little actual coercion that occurs between doctors and patients. There is undoubtedly influence; most patients feel some pressure to go along with their doctor's recommendations. We see no reason to eliminate that influence since most physicians recommend only what they believe to be in their patients' best interests. However refusing to take care of patients who would not follow one's recommendations might be considered more than influence; it might count as coercion if one were the only doctor available. Similarly, threats not to allow a patient to reenter a hospital if he leaves against the advice of his doctor may also be coercive. Not all such coercion is necessarily unjustified. Doctors coerce patients primarily for what they believe to be the patient's own good, and sometimes this paternalistic behavior may be morally justified (see Chapter 8), though often it is not. But whether justified or not, if a patient consents to a medical procedure because of coercion by the doctor or any other part of the medical team, then his consent is not valid consent.

Valid consent requires the absence of any coercion by the doctor or the medical staff. Coercion involves any threat of sufficient force that no rational person would reasonably be expected to resist it (Gert, 1972). A threat of this kind means that the person being threatened has been deprived of his freedom and so has an excuse for doing what he has been coerced to do. We do not regard strong recommendations, forcefully given, as coercive. To extend the term "coercion", so that any pressure by a doctor on a patient to accept the doctor's recommendations would count as coercive, and hence in need of justification, seems to us undesirable. We wish to allow doctors considerable leeway in supporting their views. Patients often have irrational fears that must be overcome, so we do not want to set unrealistic limits. But, given the tremendous authority of doctors and the vulnerable position of patients, no threats of lack of care should be allowed.

Psychiatrists, surprisingly enough, are probably guilty of coercing far more patients than any other doctors. This is not due to any greater perversity on the part of psychiatrists but rather to the fact that psychiatrists have a much more credible threat than other doctors. In non-psychiatric cases, the threat of refusing treatment is truly coercive only if other doctors are not available, and such cases are doubtlessly rare. However, the psychiatrist has the threat of involuntarily hospitalizing the patient, and this is not quite so rare. Given the quality of most state mental institutions, it is particularly coercive to tell a patient that if he does not consent to treatment he will have to be committed. One of the advantages (or disadvantages) of the newer, more restrictive commitment laws (see Chapter 9) is not merely that they cut down on the number of commitments. They also reduce considerably the use of the threat of commitment in order to gain consent to treatment. Again, we must emphasize that just as commitment is sometimes justified, so the threat of commitment in order to obtain consent may also sometimes be justified. However, this does not change the fact that coercion is involved and that treatment is thus being given without valid consent.

One might claim that whenever a patient has a serious risk of death or disability, his decision about treatment cannot be considered to be free. If there is only one medically acceptable treatment, this claim is true. It may thus seem that there could never be valid consent for the uniquely acceptable treatment in any serious case, but this is not so. Valid consent does not require that one be free to choose, only that there be no coercion from another person. That painful circumstances force me to accept surgery for appendicitis does not make my consent invalid. Even though, in the absence of very special beliefs, as a rational person I must accept surgery, my consent is still valid. I may not be free to choose a course of action that I believe will result in a painful death, but no rational person is interested in that freedom (except in very unusual circumstances).

The freedom that rational persons are concerned with is the freedom to choose the course of action which, on the basis of adequate information, they believe will probably result in what they judge to be the least amount of evil overall or the most favorable balance of good and evil. It is important to realize that often there is no objective "best" alternative. Even if there is complete agreement on the severity and probability of all the goods and evils involved, two persons may rationally choose different alternatives, for they may rank the goods and evils differently. Doctors often seem to rank mere survival as extremely important, so that, for them, reducing the risk of death by 10 percent outweighs considerable pain and disability. Patients often give more weight to pain and disability than to a 10 percent higher risk of death. No person rationally chooses to suffer any evil without some reason, and it is not the freedom to do so that is at question here. It is the freedom to choose for oneself, without threats and with adequate information, what one considers the lesser evil or what benefits are worth what risks.

Competence

It may seem that all that is necessary for valid consent has now been discussed: adequate information and the absence of coercion. For most persons that is correct, since most persons are competent to make their own medical decisions. But some persons are not. No matter how much information they are given and how noncoercive the situation, they cannot give valid consent because they are incompetent to do so. We mentioned at the beginning of this chapter the parallel between valid consent and a valid will. In order for a will to be valid, its maker must know, at least in general, what he is leaving and to whom, which corresponds to the condition of adequate information. Also, it cannot be made under duress, which corresponds to lack of coercion. Also, like a valid will, valid consent can be given only by someone who is competent. Thus, in order to complete our discussion of valid consent, we must give some account of the concept of competence.

Obviously, we are primarily concerned with the concept of competence because it is directly relevant to the question of valid consent. However, we believe that there is no special sense of competence which is related to valid consent. Rather, the concept of competence remains the same, though since it is always related to some particular task or group of tasks, the criteria for determining whether or not someone is competent to perform depends on the nature of the task. Thus we shall begin by analyzing the general concept of competence and then relate it to the problem of valid consent.

The sentences "John is competent" and "John is incompetent" do not express complete statements. Of course, the context may make it clear what is meant. For example, if we are discussing whether or not to hire John to design a house, it is quite clear that the sentence "John is incompetent" means that John is incompetent to design a house. But not all incompetence is attached to offices or positions or jobs. Someone can be incompetent in what might be regarded as a more fundamental sense, namely, incompetent to do some activity that almost all normal adult human beings can perform. This is not really a different sense of incompetence. Rather, the person is incompetent to do more fundamental activities than those involved in a specific office or job. For example, a person may be incompetent to feed himself. He may simply be unable to figure out what or how to eat. The incompetence shown here is the same kind involved in designing a house. In both cases, there is a specific task to be performed and a person who is unable to perform it. The only difference—a big practical difference—is that only those who occupy some position involving the designing of houses are considered incompetent when they cannot do so, whereas anyone who cannot understand how to feed himself is regarded as incompetent.

But though competence always involves the ability to do some particular task, it is not merely a synonym for ability. If someone cannot run a marathon, we do not say that he is incompetent to run a marathon; rather, we say that he lacks the physical ability to do so. Competence involves mental or volitional rather than physical abilities. But lack of such abilities does not by itself involve incompetence. If one does not have the mental ability to do theoretical physics, it would at least be misleading to say that he is incompetent to do theoretical physics, unless he were in a position in which he were expected to have that mental ability. To return to our first example, when we say that John is incompetent to design a house, this implies that John is an architect or has a job that involves designing houses. Thus, competence involves having those abilities that persons in that position are supposed to have. So that it is only with architects or engineers that lack of ability to design houses counts as incompetence.

As noted before, to say of someone that he is incompetent demands a context. A person is not simply incompetent; he is incompetent to do x, or x and y, or x, y, and z. It is possible for someone to be incompetent to do any of the things that a normal adult human being can do; newborn infants are incompetent in this total way, and so are some adults. Philosophical problems arise when someone is competent to do some kinds of things, but is incompetent to do some others. How is one to decide if such a person is competent to do some particular type of activity? The more precisely described the activity, the more likely it is that one can decide whether or not someone is competent to perform it. Suppose that we are wondering whether someone is competent to make a will. To be competent to do this, one must know what is involved in making a will; one must understand, at least in a practical sense, what a will is. If one is not aware of what is involved in making a will, then one is incompetent to make a will. In general, to be competent to do x, one must have at least a practical understanding of what it is to do x. One must also understand when one is doing x. It is not enough to know what wills are; one must also be able to understand when one is making a will. Thus two necessary features for being competent to perform an activity are that one understands what that activity is and knows when he is performing it.

We tentatively define incompetence in the following way: a person is incompetent to do x if it is reasonably expected that any person in his position, or any normal adult human being, can do x, and this person cannot (and his inability to do x is not due to a physical disability). In this account of incompetence, nothing is implied about how one ought to treat a person who is incompetent to do x, except, of course, that if one wanted to get x done, one should not entrust the job to someone who is incompetent to do it. Even with regard to someone who is incompetent to do something that every normal adult human being is expected to be able to do, nothing more is implied than that it would be unreasonable to entrust that person with the task of doing x. Thus if someone is incompetent to handle money, it would be unreasonable to give him some task that involved handling money. However, if he has money of his own, nothing follows about prohibiting him from spending it, or giving it away, or doing anything else he may want to do with it. We are not automatically justified in preventing someone from doing something simply because he is incompetent to do it. If someone is an incompetent poker player, for example, and does not know what hand beats another, it does not follow that anyone is justified in prohibiting him from playing poker if there are others who are willing to play poker with him. As we will see in Chapter 8, however, if the consequences of his playing are serious enough, one might be so justified. Similarly, if someone is incompetent to make a decision about some medical treatment, it does not follow that someone else is thereby justified in making that decision for him. It may be that someone is justified, but this does not follow solely from the fact that the person is incompetent to make the decision himself; it depends upon other matters as well.

Incompetence, in the sense that we have been discussing it, is almost completely an empirical matter and is not identical to what for the sake of clarity we shall call "legal incompetence." To be declared legally incompetent to do x depends upon a judgment of incompetence in the sense that we have been discussing. However, it also involves something else, namely, a decision that someone else is justified in acting on that person's behalf with regard to doing x, and that he may justifiably require, prohibit, or permit actions with regard to that person. Normally a judgment of legal incompetence to do x only involves incompetence to do those things that all normal adult human beings are expected to be able to do. That is, it is persons who cannot feed or clothe themselves, or who cannot handle money, whom we declare legally incompetent and then sometimes give to some other person, for example, a legal guardian, the authority to make decisions for them.

We also sometimes declare persons legally incompetent to perform certain professional tasks; however, in these cases, we do not appoint a guardian to make decisions for them. Thus, someone who has a position as a lawyer or doctor may, for various reasons, become unable to perform the professional tasks required of him. It may be appropriate in some instances for the person to be declared legally incompetent to perform those tasks. In this case, no guardian would be appointed, but the person would be prohibited from performing that activity which he performed incompetently.

The important issue in going from empirical incompetence to legal incompetence is whether one can justify the restriction of freedom involved in such a judgment. We will discuss the justification of such interference for the person's own good in Chapter 8. Here, we will only say that such a justification involves, as one of its criteria, the prevention of the suffering of significant evils.

Competence to consent to treatment

Let us now apply these general considerations regarding competence and incompetence to the problem of valid consent. What is involved in saying that someone is incompetent to give valid consent? We believe there are two levels of such incompetence and that it is useful to distinguish clearly between them.

1. There is a category of patients who are unable to give or refuse consent at all. Some patients in this category are completely unaware of their surroundings and are not able to understand any question that might be asked of them—for example, infants, patients in a coma, or patients who are severely retarded or senile. For such patients, nothing that they say or do could even count as consent or refusal of consent. They may be called "totally incompetent," and it is universally acknowledged that it is justified, even morally required, for someone else to make decisions for them and on their behalf.

However, there are other patients in this first category who are less than totally incompetent. They may have very limited cognitive abilities, may be able to ask for food, or for relief from pain, and yet be unable to understand any questions not directly related to present stimuli. Therefore, they do not understand at all the request for consent to a medical procedure. They do not know what is being asked of them and do not realize in fact that they are being asked to give consent. For these patients, as for those who are totally incompetent, it seems appropriate and morally justified for someone else to be authorized to give or withhold consent on their behalf. For example, consider the following case:

Case 3-1.

Ms. B was a 69-year-old woman with a biopsy-proven unresectable retroperitoneal sarcoma (a kind of cancer posterior to the abdominal cavity). She was admitted to the psychiatric unit in a profound state of confusion that was thought to be the result of delirium, a very severe psychotic depression, or a combination of the two. Approximately one year earlier she had been admitted to the hospital with a similar mental syndrome. At that time, a retroperitoneal mass had been identified and biopsied during exploratory surgery. Treatment with ECT at that time (one year before the present admission) resulted in dramatic clearing of her confusional state and melancholia, enabling Ms. B to resume a satisfying life with her family for a period of about ten months, when the current confusional state developed.

At this point, Ms. B was disoriented to place and time and was severely agitated and restless. She was not able to give understandable answers to most direct questions, and in general her speech consisted of incoherent babbling. An extensive search for a metabolic, pharmacologic, or structural cause for her mental syndrome yielded no positive results. Her retroperitoneal sarcoma appeared to have increased somewhat in size, but this could not be directly correlated with her change in mental function. Her sarcoma was in no way felt to be immediately life-threatening. Her physicians felt that ECT was again indicated but that she was incompetent to give even simple consent to any treatment procedure. The hospital attorney was of the opinion that ECT could be used if the unanimous consent of her three adult children were obtained. Her children did consent, and a course of ECT was again administered. A similar gratifying improvement resulted.

We will refer to both types of patients in this category as being *incompetent to give* (even) *simple consent*. The concept of "simple consent" is explained in the description of the next category of patients.

2. This second category of incompetent patients we refer to as being *incompetent to give valid consent*. They are, however, competent to give simple consent; that is, they understand that they are being asked to consent to a medical treatment and can give consent or refuse to do so,

but they lack the ability to understand or appreciate the information that is necessary to give a valid consent. The clearest example of someone who fits in this category is a patient who is moderately delirious or demented and is aware of only some aspects of his situation. He may perfectly well understand that he is being asked for consent to perform some medical procedure, but may not know where he is, who is asking for his consent, or why they are asking for it; or he may disbelieve most or all of what he is told about the consequences of his giving or refusing that consent. This person differs from the persons discussed in the first category in that he may give his consent to a treatment, or vigorously refuse to give it. But neither the refusal nor the granting of consent count as valid, for such a person does not possess sufficient information to give valid consent. We will say of such a patient that he is competent to give or refuse simple consent but is incompetent to give or refuse valid consent.

Another interesting example of patients in this category are those who have delusions which are relevant to the giving or withholding of consent. Suppose that someone has the paranoid delusion that all of his doctors are part of a plot to take over his body and that regardless of what his doctors are saying, if he gives his consent they will perform some procedure that will give them complete control over his thoughts and actions. He believes this even though consent is being requested only for a diagnostic procedure completely unrelated to his delusion, for example, a biopsy to determine if a tumor is malignant. We maintain that such a person is competent to give simply consent but is incompetent to give valid consent, because he is unable to understand or appreciate the information that is necessary for valid consent. This does not mean that we are thereby justified in performing the biopsy independent of his valid consent. For this to be true, one must apply the justification procedure that we alluded to earlier and will discuss in Chapter 8.

A patient may also have a delusion which results in his giving rather than withholding consent. Suppose a man believes that he has been given superhuman powers and that nothing done to him can harm him in the slightest way. Thus, when he is asked for his consent to undergo a serious and risky experimental operation he readily gives it, for he does not believe that there is any risk for him. In such a case, his simple consent is not valid, for he is incompetent to give valid consent. We say this for the same reason we gave above: he is unable to understand or appreciate the information necessary for valid consent.

We have shown that if someone is not given adequate information, he cannot give valid consent, for he does not know enough about that to which he is consenting. But clearly, the important matter is not merely the mechanical procedure for providing information to someone. Suppose a doctor has developed what is universally acknowledged as an ideal presentation of all the information required for valid consent in a certain medical situation. Now, suppose that he presents this information to his patient in a way that all his fellow doctors regard as clear and nonbiased. Does the patient now have the information required for valid consent? We don't know. It depends upon what the patient understands. If the patient has only a very limited command of English, then providing the information in English will not give the patient adequate information for valid consent. Similarly, if the patient is suffering from such anxiety that he can understand little of what is being told him, he does not have adequate information to give valid consent. This is why it may be important that a nurse, or someone who has the opportunity to talk to the patient at some length, be required to certify that the patient does indeed have adequate information to give valid consent.

A patient may be incompetent to give valid consent because he cannot understand anything as complex as the information required for valid consent, for example, a retarded person, a young child, or a person who may not be able to appreciate the information because, for example, he is suffering from delusions. A completely senile person and a very young infant would be incompetent to give even simple consent; the person with delusions would be incompetent to give valid consent. A slightly older child might be competent to give simple consent but incompetent to give valid consent. In general, one might say that when we can clearly determine that the person doesn't understand any of the information being provided, he is incompetent to give simple consent, and when he understands some but not enough information or doesn't appreciate it, he is only incompetent to give valid consent.

It is fairly easy to determine whether or not someone understands the information that is presented to him, though, of course, there will always be borderline cases. When dealing with appreciating the information, we have a trickier situation. As we use the term "appreciate," it requires more than understanding. For example, someone with paranoid delusions that involve his doctors may understand all of the information presented to him, but because of his false beliefs about his doctors he cannot properly evaluate that information and thus cannot give valid consent. Failure to appreciate, like failure to understand, must be determined independently of consent or refusal of consent. A refusal of consent, even if irrational, is not sufficient to show a failure to appreciate the information.

There is a significant practical difference between the two kinds of incompetence. With incompetence to give simple consent, nothing the person does counts as either giving or refusing consent. In this case, there is no overruling of the patient's decision; there is no decision of the patient's to overrule. With incompetence to give valid consent, the problem is more complex. In such cases, patients can either give or refuse simple consent, but since they are incompetent to give valid consent, obviously their consent is no more valid than their refusal. In these cases, one must decide what to do. Our suggestion is that these cases be divided into two main categories: (1) those in which the person gives simple consent but is considered incompetent to give valid consent and (2) those in which the person gives a simple refusal of consent but is also considered incompetent to give a valid refusal of consent.

In the first category, we do not include patients who actively seek a treatment that they do not adequately understand, but only those who acquiesce to their physicians' request for their consent. For these patients we believe that a guardian should be appointed to decide on behalf of the patient whether or not to accept the proposed treatment. If he agrees with the patient, there is no problem at all. If he does not agree, a problem does seem to arise, for then the guardian seems to be overruling the patient's decision. This kind of action seems paternalistic and thus in need of justification. However, this is not the case. The guardian is not overruling the patient, for the patient has simply acquiesced to his physician's request. Rather, he is refusing to make a decision that he believes is not in the patient's best interests. It is true that the patient is not getting a procedure or treatment to which he has given simple consent, but the guardian is not depriving the patient of something the patient wants; rather, he is simply refusing to provide something he doesn't think is in the patient's interest.

The second case, in which the patient gives a simple refusal of consent, is somewhat different. Again we suggest that a guardian be appointed to decide, on behalf of the patient, whether or not to accept the proposed treatment. If he agrees with the patient concerning refusal of consent, again there is no problem. However, if he does not agree and thinks that the proposed treatment is in the patient's best interest, then there is a serious moral problem. To allow the guardian to consent to treatment when the patient has refused it, even though by hypothesis the refusal is due to a failure to appreciate the situation, is to allow one person to act paternalistically with regard to another, simply on the grounds that the first person is incompetent to give valid consent. We do not think that, in and of themselves, these are adequate grounds. We believe that even with regard to patients incompetent to give a valid refusal of consent, their simple refusal of consent must be taken very seriously and overruled only in special circumstances, when the failure to treat would result in significant evils being suffered. It is an act of paternalism and has to be justified like any other paternalistic act (see Chapter 8).

We obviously think that just as with competent patients, so with those incompetent to give valid consent it is a much more serious matter to treat without consent of the patient than not to treat even though simple consent has been given. In the former case, we actively impose something on the patient; in the latter case, we simply refuse to do something agreed to by the patient. Thus, we give a simple refusal of consent for treatment, by a patient incompetent to give a valid refusal of consent, much more weight than a simple consent by the same patient. This seems to reverse the traditional practice whereby a simply consent by such a patient is sometimes taken to be valid, and a simple refusal of consent is all too easily overruled.

The competence of psychiatric patients to consent to treatment

Some might think that patients suffering from mental maladies are, in general, incompetent to give or refuse to give valid consent to proposed treatments. However, we believe that the overwhelming majority of such patients are quite competent to give valid consent and that one is much more apt to find incompetence to give valid consent among patients found, for example, on neurosurgery or oncology wards.

The case of Ms. B discussed above does represent a patient who was incompetent to give even simple consent to treatment because of her thorough confusion and incoherence. One does encounter such patients in psychiatry who, because of drug delirium or acute severe psychosis, for example, are unable to give or refuse even simple consent. Though such cases are dramatic, they constitute only a small percentage of psychiatric cases. We believe the incidence of this kind of incompetence is much higher in selected medical and surgical patient populations. This would be an interesting and important topic for future empirical research. At any rate, when one does encounter a patient who is incompetent to give or refuse simple consent, then it is usually morally acceptable to rely on the next of kin or a court-appointed guardian for consent, whether the patient is suffering from a mental or a physical malady.

We believe that the overwhelming majority of psychiatric patients are competent to give valid consent. Some patients have delusions directly related to the treatment or consent process, but their absolute numbers are very small. One encounters somewhat more frequently psychiatric patients whose mental confusion, associated with a psychosis, makes their understanding and appreciation of the consent information sufficiently suspect that one questions whether the consent they give or refuse is valid. But most patients know that something is wrong with them and are capable of understanding the nature of the available treatments and the risks and benefits associated with each. There is nothing inherently more suspect about a patient consenting to a treatment to relieve mental pain than about a patient consenting to a treatment to relieve physical pain.

When psychiatric patients do refuse treatment, whether or not they are competent to give valid consent, that refusal is frequently seen as irrational by the treatment team. We propose that irrational refusal of treatment never be taken as a sign of incompetence to give valid consent if the patient's giving of consent would have been regarded as valid. (There is usually no irrational consent, for doctors would not propose a treatment that was irrational.) We should be able to determine incompetence to give valid consent independently of the giving or refusing of consent; irrational refusal obviously can not be so determined. It is sometimes morally justified to overrule valid but irrational refusal of treatment (see Chapter 8), but one should be clear that this does not prove incompetence. One advantage of being clear about this distinction is that a change of mind (perhaps due to the persuasive abilities of the psychiatrist) by a competent patient who has previously validly but irrationally refused treatment can now be taken as valid consent; whereas if the patient had been determined to be incompetent to refuse valid consent, his subsequent consent would not be valid.

References

Cobbs v. *Grant*, 8 Cal. 3d 229, 502 p. 2d 1, 104 Cal. Rptr. 505 (1972).

Culver, Charles M., Ferrell, Richard B., and Green, Ronald M. ECT and special problems of informed consent. *American Journal of Psychiatry*, 1980, *137*, 586–591.

Culver, Charles M., and Gert, Bernard. Regionalization of surgical services (letter to the editor). *New England Journal of Medicine*, 1980, *302*, 1034–1035.

Gert, Bernard. Coercion and freedom. In Pennock, J. Roland and Chapman, John W. (eds.), *Coercion. Nomos XIV*. Chicago: Aldine-Atherton, 1972, pp. 30–48.

Luft, Harold S., Bunker, John P., and Enthoven, Alain C. Should operations be regionalized? The empirical relation between surgical volume and mortality. *New England Journal of Medicine*, 1979, *301*, 1364–1369.

Wennberg, John E. Factors governing utilization of hospital services. *Hospital Practice*, 1979, 115–127.

Wennberg, John E., Bunker, John P., and Barnes, Benjamin. The need for assessing the outcome of common medical practices. *Annual Review of Public Health*, 1980, *1*, 277–295.

4
Maladies

Introduction

How should concepts like "disease" and "illness" be defined? This is an important problem in medicine and psychiatry because of its inherent conceptual interest and because it is related to many practical matters.[1] For example, it might be claimed that disease concepts are defined quite subjectively and thus idiosyncratically by different persons, groups, and cultures. On this view there could be no objective guidelines to help decide, say, whether such conditions as tobacco addiction or having had a breast amputated represent diseases or maladies, and thus whether smoking clinics or breast reconstruction surgery ought to be covered by health insurance. On the other hand, it might be asserted that there are fairly clear and objective universal criteria for defining disease concepts. Then it might follow that the disease or malady status of most human conditions could be determined fairly clearly, and that these criteria could also help explain why a few conditions do have a genuine borderline status.

There can also be disagreements among those who believe that objective disease criteria exist, and these disagreements also have important implications. For example, if one believes that a necessary criterion for disease is the presence of some "abnormality" in physiological functioning, then mental conditions such as phobias and compulsions will likely be excluded at the outset. By contrast, if one claims that "abnormal" physiological functioning is not even a criterion for determining whether physical conditions are considered diseases, then it will be an open question whether and to what extent various mental conditions satisfy the actual criteria for disease or malady status.

We believe that objective definitional criteria do exist and that the same criteria apply to mental conditions as to physical conditions. That is, some physical and mental conditions will satisfy these criteria and thus qualify for malady status, while other physical and mental conditions will not. Thus, we differ from those who hold that mental illness either does not exist at all or else belongs in some fundamentally different category from physical illness. In this chapter, however, we will limit our discussion largely to physical conditions in order to show clearly that even when so restricted, the disease/malady criteria which emerge contain no necessary reference to bodily states or processes. In the next chapter, we will show that these criteria apply in the same way and with no less force to mental conditions. Before considering these criteria, however, it is necessary to discuss the array of malady-terms that exist in our language.

In the English language a cluster of words are used to refer to the conditions that concern us here. Three of the most important are "disease," "illness," and "injury," but there are many more: "wound," "disorder," "defect," "affliction," "lesion," and "disfigurement," to list a few. While these terms have distinct though partly overlapping connotations, which can be fairly precisely identified, there is nevertheless an arbitrary element in the labeling of the various conditions. For example, most of those conditions regarded as injuries (e.g., broken limbs) are not called diseases, and vice versa (e.g., measles is not an injury), but it seems impossible to construct any definition of disease that would not include injuries. Indeed, an occasional pathology textbook (Peery and Miller, 1971) does use the term "traumatic diseases," which is easily comprehensible despite its lack of ordinary usage. An interesting example of the arbitrary nature of this labeling is the condition experienced by deep sea divers who return from the depths too quickly. It is referred to as either "caisson *disease*" or "decompression *illness*," while essentially all of the associated ill effects are due to the cellular *injury* caused by nitrogen bubbles forming in various bodily tissues.

Illness and disease are closely related, but diseases are more robust ontologically than illnesses. They are regarded as entities having characteristic signs and symptoms with known or discoverable underlying "mechanisms" and, ultimately, known or discoverable etiologies. Diseases can exist antecedent to the appearance of symptoms (e.g., a cervical cancer detected by a Pap smear), though they nearly always lead to the suffering of symptoms. In illnesses, by contrast, symptoms are predominant and the underlying pathogenesis is almost ignored; for example, speaking of a "disease process" seems appropriate, but "illness process" sounds forced. Thus, diseases at a presymptomatic stage (whether cancer or tooth decay) would not generally be regarded as illnesses. And someone who was poisoned (e.g., by an overdose of aspirin) would generally be regarded as being ill but not as having a disease. But there seems no logical basis for this distinction; some infections (which are prototypical examples of diseases) exert their deleterious effects chiefly through the secretion of poisons (toxins).[2]

Many important kinds of conditions do not seem to be either diseases, illnesses, or injuries. For example, neither an ordinary tension headache nor a hernia seems to fit very comfortably into the disease or illness categories. A hernia seems more like an injury, but that word does not seem exactly right either. Idiopathic mental retardation does not fit any of these three categories. Nor is it clear how to classify someone who is having an allergic reaction. Even less clear is what to say of someone who has a significant allergy but is presently free of any symptoms.

It would be very useful to have a general term which includes disease, illness, injury, headache, allergy, and so on. We believe that all illnesses, injuries, diseases, headaches, hernias, and even asymptomatic allergies do have something in common. We propose "malady" as the general term that includes them all.

By "malady" we mean roughly any condition in which there is something wrong with the person. This is what all the various conditions have in common. Later, we will offer a more precise definition, but first some other definitions of "disease" will be examined to serve as a background for our own.

Some past definitions of "disease"

There have been many attempts to set out a formal definition of "disease," but among physicians only rarely has a distinction among "disease," "illness," "injury," and related terms been recognized. Consequently "illness" is often used interchangeably with "disease," and injuries are regarded merely as a subclass of diseases. What most of the following authors have intended to define is what we call a malady, so that in addition to diseases they have included injuries, illnesses, headaches, and the like in their definitions. By not realizing what they have done, however, they have often been misled by some particular feature of the term "disease."

Some definitions are general, vague, and too inclusive. Thus, a pathology textbook (Peery and Miller, 1971) states:

> Disease is any disturbance of the structure or function of the body or any of its parts; an imbalance between the individual and his environment; a lack of perfect health.

This seems to offer three separate but equivalent definitions. According to the first definition, clipping nails and puberty are diseases, as well as asymptomatic situs inversus (right-left reversal of position of bodily organs) and being tied to a chair. The second definition is too vague to be of any use, and the third is circular. A textbook on internal medicine (Talso and Remenchik, 1968) says:

> . . . disease may be defined as deprivation or lack of ease, a discomfort or an annoyance, or a morbid condition of the body or of some organ or part thereof.

There are similar problems here. Two separate definitions are offered. The first is too sweeping and would include inflation, quarrelsome in-laws, poor television reception, and ill-fitting shoes, while the second is circular and depends on the phrase "morbid condition," which is merely a synonym for "disease."

Most texts on medicine and pathology make no attempt to define disease, which is appropriate enough since the exercise is irrelevant to their purposes. Thus, it might be said in defense of the above attempts that they were probably not meant to be taken very seriously.

One cluster of definitions which *is* meant to be taken seriously utilizes a dynamic metaphor in which a person is pictured as constantly interacting and adapting to changes in his environment; disease then corresponds to a failure in that adaptation. An early expression of this definition is found in White's 1926 book, *The Meaning of Disease*:

> Disease can only be that state of the organism that for the time being, at least, is fighting a losing game whether the battle be with temperature, water, micro-organisms, disappointment or what not. In any instance, it may be visualized as the reaction of the organism to some sort of energy impact, addition or deprivation.

Thus, one wrestler held down by another is suffering a disease. A more modern version is found in Engel's 1962 book, *Psychological Development in Health and Disease*:

> When adaptation or adjustment fail and the pre-existing dynamic steady state is disrupted, then a state of disease may be said to exist until a new balance is restored which may again permit the effective interaction with the environment.

Aside from the ambiguities and the question begging inherent in "adaptation" and "balance," the emphasis in this definition (and the preceding one) on the deterioration of a previously more normal state seems to rule out all congenital and hereditary diseases. (In the next chapter, we will discuss an earlier and more precise definition offered by Engel.)

J. G. Scadding (1959, 1963, 1967) has written a series of papers on the definition of disease. In the most recent of these, he offers the following "formal definition":

> A disease is the sum of the abnormal phenomena displayed by a group of living organisms in association with a specified common characteristic or set of characteristics by which they differ from the norm for their species in such a way as to place them at a biological disadvantage.

Scadding's definition is an improvement over those previously cited. He explicitly introduces the notion of deviation from a norm for a species which, though it does not deserve the prominence he gives it, is necessary for understanding the essential elements of the concept of malady. His "biological disadvantage" criterion also points in the right direction but is too vague. Kendell, in a recent paper (1975), interprets "biological disadvantage" as meaning decreased fertility or longevity. But this does not work. Even Kendell recognizes that his revision of Scadding's definition leaves one with the "rather disconcerting" result that a condition such as psoriasis would not qualify as a disease.

Several recent authors have also correctly identified aspects of disease. Spitzer and Endicott (1978) include in their definition of "a medical disorder" that it is intrinsically associated with distress, disability, or certain types of disadvantage. We think this definition is on the right track. Goodwin and Guze (1979) consider as a disease "any condition associated with discomfort, pain, disability, death, or an increased liability to these states, regarded by physicians and the public as properly the responsibility of the medical profession." We agree with the first part of their definition but will show below that the second part ("regarded by physicians . . .") is not necessary. Many of these points were anticipated by King (1954) in a fine older paper, in which he incorporates both the notion of various evils and the deviation from a norm in his discussion of disease.

The definition of "malady"

Having something wrong: maladies and evils

We do not regard ourselves as presenting an entirely new account of "disease" or "malady"; our goal is to provide a more precise and systematic account of maladies than has been given before. Doing so will clarify the logic of the concept and thus enable more fruitful discussion of controversial cases, such as allergies. We also hope to help resolve the more general controversy about whether what is considered a malady is culturally determined or universal.

In a sense, we will determine what a malady is by seeing what it means for a person to have something wrong with himself. Briefly, to have something wrong with oneself is to have a condition, other than one's rational beliefs and desires,[3] such that one is suffering or has an increased probability of suffering some evil. Further, this condition must be a condition of oneself as an individual such that, whatever its original cause, it is now part of oneself, and hence cannot be removed simply by changing one's physical or social environment.

In this section, we shall focus on that aspect of maladies that links them intimately with the suffering of evils. It is important to note that as we use the term "evil," it has no moral connotations; one could, without changing the meaning, substitute the term "harm" for "evil," as in "make sure no harm comes to him." Many accounts of disease have linked the concept very closely to the suffering of death, pain, or disability, but without attempting to draw out the apparently obvious but unexplained common feature in death, pain, and disability.

It may seem odd to ask, "What do death, pain, and disability have in common?", but in fact it is an important question. Part of the problem is that they seem so obviously to go together that no attention has been paid to explaining what they have in common.

Apples, peaches, pears, and plums are all fruits, but what is the genus of which death, pain, and disability are species? The most salient feature they have in common is that no one wants them. In fact, everyone wants to avoid them. Thus, what unites death, pain, and disability is the attitude that people take toward them. But we cannot simply say that all persons try to avoid death, pain, and disability. There are times when persons actually seek death and willingly endure pain or disability. Generally this happens when something else has gone wrong. For example, persons seek death when life has become too painful, or they endure pain or disability to save their own lives or the lives of those they love. Thus, a qualification must be added: all persons avoid death, pain, and disability *unless they have an adequate reason not to avoid them.*

As noted in Chapter 2, we use the term "reasons" to refer to conscious beliefs that can be used to justify, that is, make rational, actions that would otherwise be irrational. We have already indicated some things that will count as reasons, for example, beliefs that someone will avoid death, pain, and disability. But these do not exhaust the beliefs that count as reasons. Beliefs that someone will avoid loss of freedom, opportunity, or pleasure are also reasons, as are beliefs that someone will gain increased abilities, freedom, opportunity, or pleasure. Note that not all reasons are egoistic; beliefs that others will avoid death, pain, and disability are also reasons.[4]

It now seems as if death, pain, and disability share the feature that all persons will avoid them unless they have a reason not to do so. But even this is not completely accurate, for some persons sometimes act irrationally and do not avoid, indeed sometimes even seek, death, pain, or disability for no reason, for example, severely depressed persons. We do not deny there is a psychodynamic explanation for their irrational actions, only that they have any conscious beliefs about benefiting anyone so that their harming themselves is a rational act. We must say then that death, pain, and disability are always avoided by persons acting rationally unless they have a reason not to do so. Though somewhat cumbersome, this description is a sufficient account of what death, pain, and disability have in common and what distinguishes them from almost everything else, for example, books, chemistry, and love. However, we can eliminate this cumbersome description by simply defining an evil (or harm) as "that which all persons acting rationally will avoid unless they have an adequate reason not to." We can now say, as we actually did at the beginning of this discussion, that death, pain, and disability are evils. Evils are the genus of which death, pain, and disability are species. What death, pain, and disability have in common is normative, and, since maladies include the suffering of evils, this explains why "malady" is a normative term.

In saying this, we mean that it is not a matter of indifference whether we or someone we care for has a malady; maladies are regarded as bad things, that is, things to be avoided. But in maintaining that "malady" is a normative term, we are not holding the view that what is considered a malady is culturally determined in a parochial sense, that is, that each culture determines its own unique set of maladies. We hold that malady is a universal concept. Death, pain, and disability are regarded as evils or harms in every society. In no society do any rational persons seek these things; rather, they all avoid them unless they have an adequate reason not to. In particular circumstances, many societies provide reasons for seeking death, pain, or disability through religious or cultural beliefs about the benefit to be gained by suffering such an evil. But anthropologists studying a culture always require some explanation for any practice that involves killing, causing pain to, or disabling oneself. They do not require explanations for saving one's own life, relieving one's own pain, or preventing disabilities to oneself. It is not a culturally determined matter that death, pain, and disability are regarded as bad things; this is a universal feature of humankind.

A malady is a condition of the person that involves suffering or the increased risk of suffering an evil. Different societies may not know that a particular condition is a malady, because it may be so endemic that they regard it as a normal feature of the species, but they can be mistaken about this just as they can be mistaken about any other matter of fact. (This is why it cannot be a part of the definition of "malady" that it be a condition regarded as the responsibility of the medical profession.) Similarly, some societies may consider a given malady as a good thing, perhaps because they regard it as a sign of the favor of the gods. This does not conflict with our analysis. Someone may have a good reason for having a malady, but it is still a malady. For example, it may be to a soldier's advantage to be sick when his cohorts are being chosen for a particularly dangerous assignment. Considered simply by itself, a malady is normally avoided; we do not deny that there are circumstances in which it can come to be regarded as something good. In classical philosophical terminology, one can say that though maladies are intrinsically bad, they can be instrumentally good.

Death, pain, and disability are not the only basic evils. Loss of freedom or opportunity and loss of pleasure must be added. By pain we mean not only physical pain but also the unpleasant feelings of anxiety, sadness, and displeasure in their various manifestations. Similarly, disabilities should not be limited to physical disabilities, but should also include cognitive disabilities, such as aphasia, and volitional disabilities, such as compulsions and phobias. All the items on this list are basic evils; every person acting rationally will avoid them unless he has an adequate reason not to. Thus, the list of evils is not arbitrary but one in which all members share a common feature. It must be stressed that the concept of an evil is not one that is simply developed ad hoc to account for the concept of a malady.[5]

Having something wrong with oneself: maladies and the absence of distinct sustaining causes

Not everything that causes or increases the risk of suffering an evil is a malady. One is suffering a malady if and only if the evil (or increased risk thereof) one is suffering is not in continuing dependence upon causes clearly distinct from oneself.[6] We will use the notion of "sustaining cause" (roughly as used by Aquinas in his Third Proof) to refer to a cause whose effects come and go simultaneously (or nearly so) with its respective presence and absence. Thus, a wrestler's hammerlock on a person may be painful, but it is not a malady. Similarly, though being in a runaway car significantly increases one's risk of incurring pain, disability, and death, one is clearly not suffering a malady, though one soon may be. A smoke-filled room can cause labored breathing and increase the risk of suffocation; extreme cold can cause pain and the loss of ability to move one's limbs easily. Nevertheless, these conditions are not maladies. However, if these situations affect a person so that he continues to suffer the evils, or is at increased risk of suffering them even when these special situations are no longer present, then that person has a malady.

Thus a person suffering a malady must have a condition not sustained by something distinct from himself. The condition may have been caused originally by factors distinct from the person, but it is not now in a state of continuing dependence on those factors; rather, it is present even in their absence. We could say that to be a malady, the evil-producing condition must be part of the person. However, for reasons of conceptual rigor, a more formal negative statement is preferable: the person has a malady if and only if the evil he is suffering does *not* have a sustaining cause which is clearly distinct from the person.

Admittedly, distinct sustaining causes cannot be determined without some residual vagueness. Precise simultaneity of cause with its effect and of deletion of cause with deletion of its effect are rare. Effects often have slight time lags: a suffocating person might not immediately recover upon receiving sufficient air, or a person in a grossly overheated room might still feel overheated after he leaves the room. Nevertheless, in principle the distinction can be made, and practice falls in line. If the time lag before recovery is brief, we consider the person as not having a malady; on the other hand, if it is highly probable that a malady will develop from a current distinct sustaining cause, we may even respond as though the malady already existed.

Our account of a clearly distinct sustaining cause may suggest that those causes are always external elements, that is, not inside one's body. However, an increased risk of evil may have a sustaining cause which, though "clearly distinct" from the person, is nevertheless *within* the person, for example, a poison capsule held in the mouth. What if the capsule is swallowed but is still undissolved? At what point does a "clearly distinct" sustaining cause become *not* so clearly distinct from the person? For example, suppose a man were dying from poison which had been dissolved, absorbed, and spread throughout his body but could be quickly removed by some chelating agent which would rapidly return him to normal. Would that make the poison a "clearly distinct" sustaining cause and consequently poisoning not a malady? If so, the next step might be to consider viruses and bacteria in the same way, namely, as sustaining causes distinct from the person, the removal of which would lead to immediate cure. But that would be absurd, since the presence of these harmful organisms in a person paradigmatically constitutes a malady. When poison, viruses, germs, and the like have become biologically integrated or cannot be removed quickly and easily, then we regard them not as distinct sustaining causes but as part of the person.

When the evil-causing element is within the person, two distinct factors determine whether the person has a malady. One is whether the element is biologically integrated; the other is whether it can be removed easily and quickly without special skill or equipment. If it is either biologically integrated or not easily and quickly removable, then the person has a malady. When integration or assimilation of a substance takes place, we always consider it a part of the person. Body cells are invaded and interacted with, injury is done, biochemical exchanges take place, and body defenses react. On the other hand, it may seem counterintuitive to regard an unassimilated foreign substance (e.g., a small marble) as part of the person. Many such items are swallowed, do not actually penetrate the body (i.e., do not traverse the gut epithelium), and are excreted with no harm done. However, if the object is causing harm and if its total removal without subsequent detriment to the person is not possible without special skill or equipment, the person has a malady (e.g., a patient who, two days after surgery, is discovered to have a clamp mistakenly left inside him).

Of course, certain vaguenesses remain because "easily and quickly removable" and "special skill or equipment" are not precise expressions and may sometimes be subject to cultural differences. Nevertheless, they give definite and helpful guidance in making distinctions at an interface which is otherwise confused. Thus, a tongue depressor or a proctoscope, though causing pain, would be clearly distinct from the person and hence not part of a malady. But a person unable to breathe because of food lodged in his trachea would have a malady, since the food is not easily removable unless someone is immediately at hand who knows the Heimlich maneuver.

These matters have always caused labeling problems. Marbles and clamps, unless they are presently causing some evil, are ordinarily simply called "accidents," without reference to disease, malfunctioning, pathology, lesion, or any of the other malady labels. However, in persisting to the fine points of what constitutes a malady, we are making distinctions at a level not ordinarily recognized or considered, so our ordinary sense of the language does not give us sufficient direction.

The role of "abnormality" in the definition of maladies

Abormality and distinct sustaining causes

It is often claimed that one is suffering a malady if one is abnormal in either structure or function. We believe that abormality, though crucial, is relevant to the concept of malady only indirectly, as a necessary feature in explaining disability, increased risk, and distinct sustaining cause.[7] When a person is suffering an evil, we decide whether he has a malady by determining if it is characteristic of members of the species to suffer a similar evil or increased risk thereof in this environment or circumstance. For example, if a woman is suffering because she has walked into a hot boiler room, she does not have a malady; rather, the environment is abnormal and should be regarded as the cause of her suffering. However, if a woman is in moderate discomfort because a cat to which she is allergic has walked into the room, then she has a malady because it is she who is, at least in this respect, abnormal. Her allergy is the cause of her suffering. Thus abnormality, though important, is relevant to maladies but only indirectly, by sometimes helping us to determine whether to assign responsibility primarily to the person or to the environment for the person's suffering. In our use of "cause," which is the ordinary use rather than the scientific, we regard the abnormal element as the cause. It is important to keep in mind that by environment and circumstances we mean those features of the situation which are distinct from the person. It is normal for all persons who have digested a significant amount of cyanide to suffer and be at increased risk of suffering evils, but that does not mean they do not have a malady. The cyanide is no longer distinct from them, hence their suffering has no distinct sustaining cause.

The description of a person's environment or circumstances can be complex, and here too the concept of abnormality can play a role. Someone is not in normal circumstances if he has not eaten for several days. A person who is suffering in these circumstances does not necessarily have a malady. We would determine whether the person is suffering from a malady by seeing whether changing the circumstances would rather quickly remove the suffering. Thus, if after eating a meal, the person who had not eaten for several days were no longer suffering any significant ill effects, then we would regard the sustaining cause of the evils to be not the state of the person but the circumstances that he was in. However, if a meal did not clear up the problems, then we would regard the circumstances as having caused a change in the person, so that the evils suffered no longer had a distinct sustaining cause. We would then say that the person had a malady, for example, malnutrition.

Abnormality and disability

The concept of disability also depends on abnormality more than it might at first seem. Is the lack of ability to walk a disability? The obvious answer seems to be yes. But if that answer is accepted, then all infants in their first year of life are disabled. In order to avoid this conclusion, the concept of normality must be used to distinguish disabilities from inabilities. This distinction allows us to say that infants suffer from an inability to walk rather than a disability. Similarly, the lack of ability to fly is not a disability in humans but an inability. Both inabilities and disabilities involve the lack of abilities, but a lack of ability is an inability rather than a disability if either: (1) it is characteristic of the species, or of members of the species, before they reach a certain stage of maturation, to lack that ability, for example, the lack of ability of humans to fly without mechanical aid, or the lack of ability of infants to walk;[8] or (2) it is due to the lack of some specialized training not naturally provided to all or almost all members of the species, for example, instruction in tennis or chess.

The clearest example of a disability is that of a person who once had an ability that is characteristic of the species (e.g., the ability to walk) and then lost it because of a disease or injury. But the feature which is most important in turning a lack of ability into a disability is that the ability one lacks is one which is characteristic of the species, not an ability that one had and has since lost. Thus, children born deaf or blind are born disabled and have not lost an ability that they once had. In contrast, a woman who, due to rigorous training, once had the ability to run a mile in five minutes and then lost that ability by discontinuing training would not thereby be disabled, for having this ability is not characteristic of the species. Thus, to be disabled is to lack some ability that is characteristic of the species at the appropriate level of maturation, such as the ability to walk, talk, or see.

We can also distinguish varying levels of an ability. Thus, a person may have the ability to walk, but only five or six steps at a time, the ability to speak, but only five or six words, or the ability to see, but only vague shapes and colors. Persons with such very limited abilities are usually thought of as suffering from disabilities. But if a man lacks the ability to walk a mile, is he suffering from a disability? It is clear that for this to be called a disability, it is not sufficient that some other person with no special training can walk a mile. For we know that some people have natural abilities that far outstrip those of most of the species, and we would not want to be forced to the conclusion that most of us are therefore suffering from a disability.

This question shows the close connection between the concept of disability and that of normality, or of what is characteristic of the species. Though some clear cases of disability present no problems, such as someone who cannot walk at all, in many other cases it is a matter of decision whether or not to call a low level of an ability a disability. However, that does not mean that there are no guidelines for making that decision. We know that in the absence of widespread environmentally caused injury or disease, the fact that the vast majority of the members of a species (of the appropriate gender, if relevant) lack an ability shows that such a lack is not a disability but an inability. It is somewhat less clear what percentage of nondiseased, noninjured members of a species must have ever had a specific ability in order to make that ability characteristic of the species and hence the lack of that ability a disability. Notice that the phrase "have ever had a specific ability" is used rather than "have a specific ability." We determine whether an ability is characteristic of the species not by seeing what percentage now have the ability but by seeing what percentage had it in their prime.

A simple majority (e.g., 55 percent) having had the ability to walk ten miles is not sufficient. An overwhelming majority of nondiseased, non-injured members of the species must have had that ability for it to be characteristic of the species. Assuming that the ability to walk is distributed along a normal curve, with the ability to walk a yard, two yards, and so on up to 100 miles or more representing points on the curve, then clear disabilities will be present when there is some discontinuity at the extreme lower end of the curve, and persons below that discontinuity will count as disabled. If there is no discontinuity, obviously there will be some arbitrariness in deciding where to draw a line so that those below the line count as disabled while those above it are thought of as simply having minimal abilities. Something like this is already done in general intelligence testing, where 100 denotes average intelligence; those having lesser scores, such as 70–80, are not regarded as mentally disabled (retarded) but merely as having lesser abilities, while those 69 and under are regarded as mentally disabled (retarded) in varying degrees.

We conclude that a person is suffering a disability when he lacks an ability that is characteristic of the species, or when he has an extraordinarily low degree of that ability. The distribution of the ability is obtained by determining the presence or absence of that degree of ability in all persons at any time during their prime, in the absence of environmentally caused disease or injury. Given the presence of a distribution of ability obtained in this way, someone is disabled if he is at the extreme lower end of the curve. Note that this concept must be modified by excluding as disabilities those lacks due to immaturity. However, once one has reached full maturity, there is no further relativizing of the concept of disabilities. It does not matter if 100 percent of ninety-nine-year-old persons lack the ability to run; their lack of ability is still a disability. But given our decision to consider only those at the extreme low end of the curve as disabled, aging will not normally result in disabilities until the sixties or seventies or later. This seems to us a welcome conclusion, for we do not normally regard those in their forties and fifties, with obviously lessened abilities, as being disabled.

Abnormality and increased risk

We have defined malady in such a way that a person has a malady not only when he is suffering an evil without a distinct sustaining cause but also when he is at increased risk of suffering some evil without such a cause. Without "increased risk," our account of malady would fail to include many conditions that almost everyone regards as maladies, such as high blood pressure.

As with disability, one must choose between interpreting "increased risk" as increase in risk over the previous state of the person or over what is characteristic of the species, even though in the clearest cases they go together. A person who develops high blood pressure has an increased risk of suffering an evil in both senses. Similarly, a person who has ingested dioxin, which is now stored in his fat, might not now be suffering any evil, but he has significantly increased the risk of such suffering because dieting or continued physical exertion may cause the fat to become metabolized by the body, thus allowing the poison to take its toll. The person's risk is increased both over what it was before the dioxin was ingested and over what is characteristic of the species. However, if a person is in extraordinarily good shape due to some training regimen, including both diet and exercise, and then discontinues this regimen, he may come to have a greater risk of suffering some evil than he did formerly. But though now at increased risk compared with his former state, he may not be at any greater risk of suffering an evil than members of the species typically are in their prime. Clearly, he does not have a malady. Further, someone may be born with a genetic defect which puts him at increased risk of suffering some evil compared with what is characteristic of the species, even though quite clearly it is not increased over what it was for him formerly. Such a person would be suffering a malady. Thus, "increased risk," as the phrase is used in the definition of malady, must be understood as "increased over what is characteristic of members of the species," not as increased over what it was formerly. This makes it clear that "increased risk," like "disability," depends upon a concept of abnormality. But again, abnormality does not enter directly into the definition of malady, but only indirectly by determining what counts as a disability or an increased risk.

Though our definition of malady makes no explicit mention of the severity of the evil suffered, or of the significance of the increased risk, actual usage does take these matters into account. We do distinguish between major maladies (e.g., terminal cancer, a broken leg, or severe migraine headaches), minor maladies (e.g., a mild case of German measles, a sprained thumb, or a mild tension headache), and those conditions in which the evil is so slight and transitory that one may prefer not to regard it as a malady at all (e.g., a slight reaction to a vaccine or a slightly stiff muscle due to a hard workout). There are structural deviations which are not associated with any increased risk of evil and hence are not maladies, such as asymptomatic situs inversus and absence of the appendix. Similarly, there are at least two kinds of pragmatic considerations that influence where, along the continuum of increasing risk, the label "malady" should be applied. First, how severe will the evil be if it does occur? If it will be slight, we tend to require a significantly increased risk before labeling the condition a malady. However, if it might be severe, we accept a much lower probability of occurrence in calling the condition a malady. Second, can we help prevent the evil from occurring? If early treatment would result in lessened risk or decrease the severity of the evil if it did occur, we would be much more likely to label the condition a malady.

Abnormality and the other evils: loss of pleasure and loss of freedom or opportunity

The evils (or increased risk thereof) most commonly present in maladies are death, pain, and disability. As we saw earlier, however, there are other evils: the loss of pleasure and the loss of freedom or opportunity.

Although the loss of pleasure (anhedonia) can be a distinct symptom of schizophrenia (and is distinguishable from the negative feelings of sadness or anxiety), its presence is neither necessary nor sufficient for making the diagnosis of schizophrenia. However, were there a human condition characterized solely by the loss of pleasure (perhaps secondary to a stroke affecting the limbic system), it would qualify as a malady according to our definition.

Loss of freedom or opportunity is an evil not uncommonly encountered in life, but most frequently it is due to a distinct sustaining cause. A man locked up in jail is suffering from a loss of freedom, but he does not thereby have a malady, for it is clear that his loss of freedom has a distinct sustaining cause. Similarly, someone threatened with violence unless he refrains from an action is not regarded as having a malady, for his loss of freedom has a distinct sustaining cause: the threat of violence by another. However, there are interesting cases in which the threat of harm is due not to the actions of another but to something about oneself. There is no question, for instance, that when one is suffering an allergic reaction, one has a malady. But what about those times when one is not exposed to the allergen? There may still be an increased risk of suffering an evil because of the allergy. But if it is relatively easy to avoid the allergen, it does not seem accurate to say that one is really at increased risk. However, no matter how easy it is to avoid eating fish, or fava beans, one still lacks the freedom to eat these items.

The man who moves to Arizona to be free of allergic reactions has avoided the pain and discomfort usually associated with his allergy, but he still suffers from a loss of freedom or opportunity to live in certain parts of the country and therefore has a malady.[9] Even if he has no desire to leave Arizona and thus is not bothered by his loss of freedom, he still has lost that freedom, just as a prisoner with a life sentence who wants to stay in jail has still lost his freedom.

Both abnormality and the loss of freedom and opportunity play an important role in the case of serious disfigurement. Though serious disfigurement almost always causes psychological suffering, anxiety, and/or depression, we regard someone suffering from such disfigurement as having a malady even in those very rare cases when he is not suffering psychologically. Serious disfigurement is a malady because in a normal environment one is deprived of some freedom or opportunity, even though in a special environment one may feel no actual loss. Note that simply by labeling the condition as serious disfigurement, we are making a judgment that the person's appearance is significantly outside of some norm. Since we count something as a serious disfigurement only if it would cause, at least initially, a nearly universal unfavorable reaction from others, we regard it as a malady because it limits one's freedom or opportunity.

We realize that if serious disfigurement is regarded as a malady, and that if there is no way to avoid at least some cultural differences in what is considered seriously disfiguring, this kind of malady will sometimes differ in different cultures. But even in the case of disfigurement, cultural relativity is important only in the marginal or borderline cases of maladies, not in the central or core examples. We do not think it culture-bound to hold that there is often still something wrong with seriously disfigured burn victims, even if they have recovered all of their abilities, are no longer in physical pain, and claim not to be suffering psychologically in any way from their burned condition. If such cases were to occur, we would still regard treatment of their serious disfigurement as treatment of a malady and hence believe it should be covered by medical insurance.

Another difficult case is that of someone who formerly suffered a chronic malady which is now perfectly controlled by available therapy so that he is no longer suffering or at increased risk of suffering an evil. For example, consider someone with hypothyroidism. If it were possible to implant a lifetime supply of completely safe and effective replacement hormone, should we count this person as having a malady? So far, we have few medications, artificial organs, or health aids that completely eliminate all evils and any increased risk. Thus at present almost everyone who depends upon continuing implanted medication or artificial implants will have a malady according to our definition. As technology improves, however, the situation we are now only imagining may become common. Should we say that such persons, whose evils and increased risk thereof have been eliminated by continuing treatment, do not have maladies? If the artificial aid were in the body, we would say they have no malady.[10] If they were dependent on external medication, however, we would say they still have a malady since their dependence on this external source limits their freedom.

Advantages of the definition

What are the conceptual advantages of our account of malady? One advantage is the introduction of a general word—malady—which refers to the genus of which all the related conditions (such as disease, illness, dysfunction, handicap, injury, sickness) are species. The term "malady," though more general than the commonly used related terms, is nevertheless explicit, precise, and usable.

A person has a malady if and only if he has a condition, other than his rational beliefs and desires, such that he is suffering, or at increased risk of suffering, an evil (death, pain, disability, loss of freedom or opportunity, or loss of pleasure) in the absence of a distinct sustaining cause.

The concept of malady can be universal and objective and at the same time have values as an integral part, namely, those values that are universal. Thus, our explication shows the inadequacy of the current fashion of regarding disease as being heavily determined by subjective, cultural, and ideological factors.[11] We have also tried to show the logic of cultural influences in those few instances in which they do occur.

One possible ambiguity about our claim to universality and objectivity should be clarified. Some might claim that diseases are simply human inventions or constructs, that "disease" is the way humans construe some of nature's processes which humans do not like. Nature in itself, independent of human interests, has no diseases, they would say. For them, "objective" and "universal" would seem to mean "in the nature of reality" or "true of the world apart from the human mind." Now, what the world is like apart from the existence of humans and from a human's knowing of it, we would not venture to say. But a second and more important sense of something's being objective is that there is agreement about it by all rational persons. Colors are not objective in the first sense, but they are in the second. It is in this second sense that the evils cited in our definition are objective and universal. Just as all rational persons agree on the color of most objects, they also agree that certain things are to be avoided unless one has an adequate reason not to. Both matters are universal and objective, based not on personal or subjective influences but on features common to all rational persons.

The concept of a distinct sustaining cause allows us to limit what counts as a malady to what seems intuitively accurate, namely, to those conditions of the person that are not dependent upon a continuing environmental state. Additionally, incorporating the notion of "increased risk" in our account of a malady allows us to include as maladies such conditions as high blood pressure.

Our explication is also meant to clarify the role of abnormality in the concept of malady. Abnormality is often taken as central to the concept of malady, but it actually plays a less direct role; it helps us decide what counts as a disability, an increased risk, or a distinct sustaining cause. There has been a general tendency in the medical-scientific world to establish a normal range for this or that (some element of the human body), and ipso facto to have "discovered" two new maladies—hyper- and hypo-this or that (Murphy, 1976; Bailey, Robinson, and Dawson, 1977). Our account makes it clear that this use of abnormality represents a misunderstanding of the concept of malady.

Our account of malady is based on elucidating the common elements in such conditions as disease, illness, injury, and so on. Using pain, death, disability, and so on, along with the absence of a distinct sustaining cause as necessary features of any malady, considerably lessens the influence of ideologies, politics, and self-serving goals in manipulating malady labels. We will return to this point in the next chapter's discussion of mental maladies. We do not mean to suggest that our concept of a malady is without vagueness. Indeed, we have discussed some of the difficult cases. The advance is in knowing more precisely where those gray areas are, why they exist, and what variables would have to be decided to resolve the question in particular cases.

A more subtle benefit of using "malady" in this new technical sense is that it is the first explicit term in any language with the appropriately high level of generality. No language that we have investigated (English, French, German, Russian, Chinese, or Hebrew) contains a clearly recognized genus term of which "disease" and "injury" are species terms. Each term in the usual cluster of malady terms has specific connotations which guide and significantly narrow its use. Of course, that is as it should be if specificity is desired and justified. "Disease," "injury," "illness," "dysfunction," and other such terms overlap somewhat, yet each has its own distinct connotations.

A final advantage of our explication of malady is that its basic elements, concepts, principles, and arguments are the same when applied to mental maladies. The usual bifurcation between mental and physical maladies disappears. Pain, disability, and the absence of distinct sustaining causes are applicable to the mental domain as well as to the physical.

Notes

1. This chapter is adapted from Clouser, Culver, and Gert, 1981.
2. A number of philosophers have distinguished illness from disease (Boorse, 1975; Engelhardt, 1976; Margolis, 1976), but we do not think this distinction is relevant to the point which concerns us here.
3. We exclude rational beliefs and desires from the evil-causing conditions that count as maladies because we do not think that someone has something wrong with him if he has a rational belief, for example, that his wife is dying of cancer, that causes him pain, or if he has a rational desire, for example, to go mountain climbing, that increases his risk of death and disability.
4. See Gert (1975, Chapter 2) for a fuller discussion of reasons and rational action.
5. For a fuller account of evils, see Gert (1975, Chapter 3).
6. We also exclude rational beliefs and rational desires; see note 3.
7. Thus we disagree significantly with Boorse (1977), who makes abnormality the central feature of disease.
8. There are also sexual differences which are characteristic of the species, so that it is an inability that men cannot bear children. See also Boorse (1977).

9. Thus we disagree with Spitzer and Endicott (1978), who explicitly regard such a person as not having a medical disorder (malady).
10. Metal plates surgically implanted to cover bony defects in the skull do seem to constitute the elimination of a malady. Suppose a pacemaker were developed that resulted in no increased risk. Then a defective pacemaker would be part of a malady if its malfunctioning were directly causing evils or increased risk of evils. Though this situation may seem at first counterintuitive, we suspect it is because the use of synthetic materials integrated into the human body is relatively novel, and our language has not yet adjusted. It is an advantage of our concept of malady that it covers the malfunctioning, breaking, decomposition, or clogging of pacemakers, artificial hips, synthetic materials, or transpecies arteries.
11. Cf. Sedgwick (1973).

References

Bailey, Alan, Robinson, David, and Dawson, A. M. Does Gilbert's disease exist? *Lancet*, 1977, *1*, 931–933.

Boorse, Christopher. On the distinction between disease and illness. *Philosophy and Public Affairs,* 1975, *5*, 49–68.

Boorse, Christopher. Health as a theoretical concept. *Philosophy of Science*, 1977, *44*, 542–573.

Clouser, K. Danner, Culver, Charles M., and Gert, Bernard. Malady: A new treatment of disease. *Hastings Center Report*, June, 1981, *11*, 29–37.

Engel, George L. *Psychological Development in Health and Disease*. Philadelphia: Saunders, 1962.

Engelhardt, H. Tristram. Ideology and etiology. *Journal of Medicine and Philosophy*, 1976, *1*, 256–268.

Gert, Bernard. *The Moral Rules*, 2nd ed. New York: Harper, 1975.

Goodwin, Donald W., and Guze, Samuel B. *Psychiatric Diagnosis*, 2nd ed. New York: Oxford Univ. Press, 1979.

Kendell, R. E. The concept of disease and its implications for psychiatry. *British Journal of Psychiatry*, 1975, *127*, 305–315.

King, Lester S. What is disease? *Philosophy of Science*, 1954, *21*, 193–203.

Margolis, Joseph. The concept of disease. *Journal of Medicine and Philosophy*, 1976, *1*, 238–255.

Murphy, Edmond A. *The Logic of Medicine*. Baltimore: Johns Hopkins Univ. Press, 1976.

Peery, Thomas M., and Miller, Frank N. *Pathology*, 2nd ed. Boston: Little, Brown, 1971.

Scadding, J. G. Principles of definition in medicine with special reference to chronic bronchitis and emphysema. *Lancet*, 1959, *1*, 323–325.

Scadding, J. G. Meaning of diagnostic terms in bronchopulmonary disease. *British Medical Journal*, 1963, *2*, 1425–1430.

Scadding, J. G. Diagnosis: the clinician and the computer. *Lancet*, 1967, *2*, 877–882.

Sedgwick, Peter. Illness—mental or otherwise. *Hastings Center Studies*, 1973, *1*, 19–40.

Spitzer, Robert L., and Endicott, Jean. Medical and mental disorder: Proposed definition and criteria. In Robert L. Spitzer and Donald F. Klein (eds.), *Critical Issues in Psychiatric Diagnosis*. New York: Raven, 1978, pp. 15–39.

Talso, Peter J., and Remenchik, Alexander P. *Internal Medicine*. St. Louis: C. V. Mosby, 1968.

White, William A. *The Meaning of Disease*. Baltimore: Williams and Wilkins, 1926.

5

Mental Maladies

Introduction

In the analysis of malady in the preceding chapter, we used primarily physical conditions of the person as examples. Yet our definition of malady contains no reference to bodily (biochemical or physiological) states or processes. In the case of physical maladies, we usually assume that there will be accompanying bodily alterations, but their presence or absence does not determine whether we consider a condition to be a malady.

One example of this is the condition of trigeminal neuralgia (tic douloureux). It is described as follows in *The Merck Manual* (1977):

> Bouts of severe brief lancinating pain in the distribution of one or more divisions of the 5th cranial nerve. . . . The cause is unknown, and no pathological changes can be found. . . . Pain is intense, and although each bout is brief, successive bouts may incapacitate the patient. (P. 1464)

Later, in discussing the diagnosis of the condition, the *Manual* notes:

> No signs, clinical or pathological, occur with trigeminal neuralgia, so that finding a sensory abnormality or cranial nerve dysfunction rules out trigeminal neuralgia as the cause of pain. (P. 1464)

Surely everyone would consider trigeminal neuralgia a malady, yet it is known only by the presence of a certain type of pain together with the absence of various possible bodily abnormalities. Of course, almost everyone believes there *is* some as yet undiscovered underlying pathophysiology; the limited distribution of the pain to one or another branch of the trigeminal nerve makes this seem very likely. But the condition of trigeminal neuralgia is considered a malady just on the basis of the evil (pain) suffered itself.[1]

One could argue that we unhesitatingly call trigeminal neuralgia a malady only because we do assume that an associated physiological abnormality will some day be found. But consider the case of schizophrenia. We do not fully understand its etiology. It is a condition known to us almost exclusively through a variety of psychological manifestations. Some believe that schizophrenia chiefly depends on certain inherited neurochemical and neuropathological abnormalities; others believe that particular kinds of early psychological experiences play the predominant etiological role; and still others hypothesize that both these factors are conjointly necessary to produce the condition.

Suppose there are three different etiological paths to schizophrenia. It might be that some people have inherited certain nervous system abnormalities so that schizophrenia will develop no matter what kind of early childhood they experience; other people have such psychologically traumatic early childhoods that schizophrenia will develop even if they have a normal nervous system; and still others develop schizophrenia only because they have both an inherited central nervous system abnormality and a traumatic childhood.

Suppose further that these groups are phenotypically indistinguishable in terms of the psychological symptoms they manifest, the course of their condition, and the most effective treatment for them. Imagine, however, that with sophisticated neurochemical assays we can detect enzymatic abnormalities in the first and third groups that are not present in the second group. (We assume that cerebral processes are a necessary substrate for the symptoms of the second group but that this group does not show the enzymatic abnormalities.)

If the pathogenesis of schizophrenia turns out to resemble the above model, which it very well could, then those who believe that bodily abnormalities are a necessary condition for malady status would presumably believe that only the first and third groups qualify. We maintain that if the second group manifested the same symptoms, course, response to treatment, and so on, it would be pointless to deny that it shared the same malady status.[2] The patients would suffer the same evils, would be treated by the same physicians in the same way, and should be covered by the same medical insurance. We believe the criteria actually used to determine whether a condition is a malady are those given in our definition in the preceding chapter: the suffering of evils (or increased risk thereof) in the absence of a distinct sustaining cause. We believe that schizophrenia and many other mental conditions satisfy this definition regardless of their etiology, a point to which we will return later.

Thus, we hold that schizophrenia is a malady regardless of whether some, all, or no schizophrenics are eventually proven to have inherited nervous system abnormalities. But if all schizophrenics have abnormal nervous systems, would schizophrenia be classified not as a mental malady but as a physical malady? Before discussing mental maladies further, we will analyze the difference between mental and physical maladies.

Mental maladies versus physical maladies

Mental maladies are distinguished from physical maladies primarily by the types of symptoms or evils that characterize them. Though etiology plays some role, as does the type of treatment that is effective in relieving the symptoms or curing the malady, it is the dominant symptoms that play the largest role in determining whether we regard a person as having a physical or a mental malady.

The concept of the mental is a very difficult one to make clear. There is a constant temptation to regard the mental as a special sphere of consciousness which is parallel to the physical, and while related to it is nevertheless distinct from it. Of course, we all now believe that the mental is intimately related to the brain, but physical pain and physical abilities and disabilities usually depend upon the brain as well, so that talking about the brain does not help one to distinguish physical from mental pain and disability. We suggest the following: one talks of physical pain and disability when and only when some particular part of the body, other than the brain, is necessarily involved in suffering the pain or disability. (When the brain is clearly involved, such as in a brain tumor or lesion, we usually speak of both mental and physical disabilities being present.) If the pain or disability is not restricted to a particular part or parts of the body, then we speak of it as mental pain or mental disability.

Physical maladies have as their predominant symptoms physical pain and physical disability. Both physical and mental maladies, such as cancer and depression, may involve an increased risk of death. Mental maladies are maladies in which the evil(s) being suffered are primarily mental pain and mental disability. Mental disability includes both cognitive and volitional disability.

Mental versus physical pain

Physical pain is always localized to some part of the body. The pain need not be focal—it can be extensive, as in a widespread burn—but still it is described as restricted to particular bodily locations.

Mental pain or suffering, by contrast, is experienced by the whole person; for example, depression and anxiety are states of the person and not states of part(s) of the body. Mental pain can be accompanied by unpleasant localized body states, such as anxiety with cardiac palpitations, but in such cases the mental state is recognized as being primary and the physical symptoms as being associated with the mental state. All of the dysphoric feeling states—depression, anxiety, self-loathing, and so on—are varieties of mental pain and play a central role in the characterization of many mental maladies.

Mental (cognitive and volitional) versus physical disabilities

Almost all normal human actions involve the exercise of what we shall call voluntary abilities.[3] We call these complex abilities voluntary abilities because they involve all of the simpler abilities that are needed in order to act voluntarily. The three kinds of simple abilities that make up a voluntary ability are physical, cognitive, and volitional abilities. For example, in order to perform that series of voluntary actions which we call playing tennis, one needs the physical ability to raise one's arm, the cognitive ability to understand the rules of the game, and the volitional ability to will to hit the ball. It is only when a person cannot perform some voluntary actions that we become concerned about which component of the complex voluntary ability they are lacking. Sometimes they lack the requisite ability because they have not been trained or taught, but in our discussion we shall only be concerned with those lacks of ability that are properly called disabilities. Thus we shall discuss physical, cognitive, and volitional disabilities.

We shall talk of a physical disability when and only when some particular part of the body, other than the brain, is necessarily involved in the lack of that voluntary ability. The lack of ability to walk or jump, for instance, are usually physical disabilities. This does not mean that the brain is not involved, only that, in addition, another particular part of the body is necessarily involved. The contrast is with the lack of a cognitive ability, such as the ability to add numbers. One can add numbers without necessarily using any particular part of one's body. Usually one adds numbers by using one's hands to write down the numbers, but one can also add numbers simply by saying the appropriate numbers. Thus, the lack of ability to add numbers, not due to lack of training, is regarded by us as a cognitive disability. When assessing the lack of ability to add numbers, we are usually unconcerned with any particular part of the body, such as the hands or mouth. We call it a cognitive disability rather than a physical disability because it does *not* involve any particular part of the body, not because it *does* involve some special nonphysical part of the body.

On this account, the various sensory disabilities—the lack of ability to see, hear, smell, touch, and taste—are physical disabilities, because they necessarily involve some particular part of the body. We realize that this seems an odd way to distinguish a cognitive disability from a physical disability. However, we are confident that upon examining those disabilities that people are inclined to call physical disabilities, one will note that some particular part of the body (other than the brain) is always and necessarily involved in the lack of that ability, whereas with cognitive mental disabilities no particular part of the body is necessarily involved.

Physical disabilities are usually not the province of psychiatry. But if a physician cannot find anything wrong with that part of the body which is involved in a physical disability, or anything wrong with the nervous system, he is inclined to regard it not as a genuine physical disability but as a conversion disorder. Conversion disorders (see DSM-III, 1980, pp. 244–247) can involve sensory disabilities (e.g., anesthesia or blindness) or motor disabilities (e.g., paralysis of a limb or lack of ability to speak).

There is a great variety of cognitive disabilities, such as poor memory and the lack of ability to think abstractly. Many cognitive disabilities are associated with known damage to or dysfunction of particular parts of the brain, and in these cases it seems arbitrary whether the condition is regarded as a mental or physical (neurological) malady. Such mental maladies are called "Organic Mental Disorders" and given extensive and sophisticated treatment in DSM-III (1980, pp. 101–162); they are equally appropriately described, though usually less thoroughly, in neurology textbooks.

On occasion, however, one encounters cognitive disability in the absence of demonstrable brain dysfunction; such a condition is considered to be a mental malady and primarily within the province of psychiatry. One example is that of a person who is unable to recall various kinds of important personal information but who appears to have no brain dysfunction. In such a patient the diagnosis of "Psychogenic Amnesia," a mental malady, would often be appropriate (see DSM-III, 1980, pp. 253–255).

Volitional disability

We classify volitional disabilities together with cognitive disabilities as mental disabilities because they do not involve any particular part of the body. This third kind of disability is of great importance in psychiatry. Volitional disability is an evil which is seen in a great variety of mental maladies. For example, individuals with phobias may not have the volitional ability to will to do some particular kind of action, such as entering an elevator; individuals with compulsions may not have the volitional ability to will not to perform some particular kind of action, such as washing their hands many times daily. Alcoholics frequently do not have the volitional ability to will not to drink over an extended period of time. Because of the importance of volitional disabilities to medicine and psychiatry, we will devote most of Chapter 6 to them.

The similarity between mental and physical maladies

In our account of maladies there is no fundamental difference in kind between physical and mental maladies. Both are defined by the suffering of evils (or increased risk thereof) in the absence of a distinct sustaining cause. The evils suffered in mental maladies can be just as intense and unwanted as those suffered in physical maladies, and both mental and physical maladies can be associated with an increased risk of death. Finally, it is not a necessary feature of either physical or mental maladies that any known underlying physiological abnormality exists.

Before discussing the important question of what kinds of mental conditions do and do not qualify as mental maladies, we will critically review two other proposed definitions of mental malady (disease, disorder). Both allude more or less satisfactorily to the suffering of evils but encounter serious difficulties in their other features.

Other definitions of mental maladies

DSM-III and "mental disorders"

Before discussing DSM-III's definition of "mental disorder," it is important to say something about DSM-III itself because we will frequently refer to it in this and the next chapter. DSM-III, the third edition of the

American Psychiatric Association's official *Diagnostic and Statistical Manual of Mental Disorders*, was published in 1980. It is a far more ambitious and thorough manual than the earlier DSM-II or DSM-I. It explicitly sets forth criteria for diagnosing the various mental disorders recognized by American psychiatry, and it attempts to furnish clear definitions for all of the terminology it employs. The coordinators of DSM-III successfully amalgamated the careful thinking of hundreds of contributors. It is above all else an intellectually honest work, and it will easily stand as one of the most significant books in the history of American psychiatry.

It is because DSM-III explicitly conveys its definitions and criteria that it can be scrutinized philosophically. In what follows we will sometimes be critical of DSM-III, but that it even allows such criticism is a mark of its worth; if a book like ours had been written five years ago, DSM-II (1968) would probably have never been mentioned.

DSM-III gives the following definition of a mental disorder (p. 363):

> . . . a mental disorder is conceptualized as a clinically significant behavioral or psychologic syndrome or pattern that occurs in an individual and that typically is associated with either a painful symptom (distress) or impairment in one or more important areas of functioning (disability). In addition, there is an inference that there is a behavioral, psychologic, or biologic dysfunction, and that the disturbance is not only in the relationship between the individual and society. When the disturbance is limited to a conflict between an individual and society, this may represent social deviance, which may or may not be commendable, but is not by itself a mental disorder.

Though the authors of DSM-III deny that they are offering a definition, they treat it as if it were one, even including it in the glossary, so we feel justified in considering it a definition. As such, it does refer in an appropriate way to some of the features of mental maladies but it has significant shortcomings. We will examine the DSM-III statement in terms of our own definition, which we believe to be more precise and usable.

The suffering or increased risk of suffering evils. The first sentence of the DSM-III definition does mention two of the evils seen in mental maladies. No mention is made of the evil of increased risk of death, which is an important evil in characterizing such maladies as depression and anorexia nervosa. The first of the evils which DSM-III does mention ["a painful symptom (distress)"] corresponds to our concept of mental pain.

But what DSM-III defines as a disability ("impairment in one or more important areas of functioning") is broader than our definition of disability and does not distinguish between a disability and an inability. For example, in both accounts a person who cannot read despite having had a good education has a disability and has a malady; he might, for example, be given the diagnosis of Developmental Reading Disorder (DSM-III, 1980, pp. 93–94). However, though we both agree that a person who cannot read because of lack of education does not have a malady, DSM-III would say that he had a disability (an impairment in an important area of functioning) whereas we would say that he had an inability rather than a disability. It is in order to prevent this lack of ability from counting as a malady that DSM-III must add to their definition, "an inference that there is a behavioral, psychologic, or biologic dysfunction." But this inferred dysfunction creates problems for the definition which we will discuss below.

The DSM-III definition also claims that the evils are only "typically" associated with mental disorders. It is unclear what is meant by this. "Typically" could mean that a person may have a malady who is not yet suffering any evils but is only at increased risk of suffering them. If so, then we have no serious disagreement, for the concept of increased risk is a necessary element in defining mental as well as physical maladies. For example, unless increased risk is included, many manic episodes would not qualify as mental maladies at all, since these episodes may for a time be unaccompanied by the suffering of any evils.

It is also possible that "typically" appears in the definition because the authors of DSM-III wanted to classify as sexual disorders many conditions which do not involve either the suffering of evils or an increased risk of such suffering, a subject we discuss in more detail below.

The absence of a distinct sustaining cause. In our definition, this feature is included because of the necessity of specifying that in maladies, the symptoms must be due to something wrong with the person rather than something abnormal in the person's environment. The DSM-III definition attempts to capture this same notion by stating that one infers that the person has some behavioral, psychologic, or biologic dysfunction and the disturbance is not limited to a conflict between an individual and society.

There are two problems with this part of the DSM-III definition. The less serious one is that nowhere does DSM-III define what is meant by dysfunction. The notion of biologic dysfunction is reasonably clear when applied to organs of the body, such as the liver, kidney, or heart. Psychologic dysfunction is a more problematic concept, but a fairly clear account might be given of what it means to infer that a psychologic dysfunction is present when one is confronted with someone who is suffering evils. But what is a behavioral dysfunction? Of what disorder would one infer that a behavioral dysfunction is present? Some disorders, like phobias and compulsions, which we call disabilities, might be said to *be* behavioral dysfunctions, but here there seems to be no useful distinction between a disability and a dysfunction; the disability is the dysfunction. Thus, to say of behavioral disorders that there must also be an inferred dysfunction is, at the very least, misleading. Part of the problem is that the concept of dysfunction needs considerable elucidation in order to do the work assigned to it by the authors of DSM-III.

But the more serious problem with this part of the definition is that we are simply told *that* an inference is made, not how or why it is made. When a person is suffering "distress" or "disability," how or why *do* the authors infer that some dysfunction is present? We believe that the inference is made when these (and other) evils are being suffered in the absence of a distinct sustaining cause, but the DSM-III definition does not incorporate this concept. Therefore, it can only state that a dysfunction is inferred to be present, while skipping over the more basic question of explaining how or why one is tempted to make such an inference.

We suspect, from the last part of their definition, that the DSM-III authors were cognizant of the need for the concept of the absence of a distinct sustaining cause. They emphasize that if the person's "disturbance" arises only in the relationship between the person and society, it is not a mental disorder. They make this point to guard against the possible political abuse of psychiatry by governments. Acts of political dissidence might otherwise be labeled mental maladies. But we think that our concept of the absence of a distinct sustaining cause, which is both more general and more precise than "disturbance limited to a conflict between an individual and society," accomplishes this same goal in a less ad hoc fashion. If a person is suffering or at increased risk of suffering evils principally because of conflict with his social environment, then his social environment would be a distinct sustaining cause of his suffering and he would not have a malady. And of course, many social dissidents or deviants do not suffer any significant evils.

Thus, the DSM-III definition correctly alludes to the suffering of evils but does not characterize these evils with sufficient precision; further, its lack of the concept of the absence of a distinct sustaining cause is not adequately balanced by its reference to inferred dysfunction, and to a disturbance which is not limited to a conflict between an individual and society.

Engel: grief as disease

In a well-known article, George Engel (1961) argues that grief is a disease. He does this by giving a definition of disease and then showing that grief satisfies each of the elements of his definition. Engel's definition is apparently meant to apply to all disease, physical as well as mental, but we examine it here because it is a mental condition, grief, which he discusses.

Engel argues that grief is a disease because it has the same characteristics as other diseases:

> . . . it involves suffering and an impairment of the capacity to function, which may last for days, weeks, and even months. We can identify a constant etiologic factor, namely, real, threatened, or even fantasied object loss. It fulfills all the criteria of a discrete syndrome, with relatively predictable symptomatology and course. (P. 18)

In effect, Engel is giving three features—(1) suffering and an impairment of the capacity to function, (2) constant etiology, and (3) predictable course—which, taken together, qualify a condition as a disease. He shows elsewhere in his article that grief does have each of these three features. (Note that Engel's characterization of a disease here is more precise than the adaptational model found is his 1962 book, cited in Chapter 4.)

However, these three features are not sufficient to make a condition a disease. The first ("suffering and an impairment of the capacity to function") does allude to the necessary condition of suffering an evil. It is not exactly clear what Engel means by "impairment of the capacity to function," but the phrase seems to connote more of what we define as a disability than does DSM-III's "impairment in one or more important areas of functioning." Engel, like the DSM-III authors, does not mention the evil of risk of death or increased risk of any of the evils. The third feature ("a discrete syndrome with relatively predictable symptomatology and course") is commonly found in diseases but is not a necessary characteristic. One can easily imagine the first appearance of a new infectious or toxic disease in which the nature of the developing syndrome, its symptomatology, and its course were all unknown and yet its status as a disease was immediately accepted by everyone.

It is the second feature (the identification of a "constant etiologic factor") that is most crucial in assessing Engel's definition. Engel's characterization of the constant etiologic factor is incomplete. Some constant etiologic factors are distinct sustaining causes, while others are not, and the difference is crucial in determining whether a condition qualifies as a malady. For example, a man awaiting execution on death row would satisfy all of Engel's criteria: he would be suffering; he would be unable to function in a variety of ways; there would be a clear constant etiologic factor; and his fear response would consist of a well-known syndrome of psychophysiological signs and symptoms with a relatively predictable course. By our definition, however, he would not be suffering a malady because his situation of being on death row, though a constant etiologic factor, would be a distinct sustaining cause of his suffering. Thus, "constant etiologic factor" is too broad a concept to be useful in defining maladies.

We believe that Engel's lack of the concept of a distinct sustaining cause leads him astray in classifying grief as a disease. Is the etiologic factor in grief a distinct sustaining cause? If a man is grieving over the death of his wife (with what DSM-III calls "uncomplicated bereavement"; see below), is her death a distinct sustaining cause of his suffering? We believe it is. If he learned that he was mistaken about her death (e.g., she is found alive a week after a remote plane crash which was believed to have killed her and everyone aboard), his grief would rather quickly cease. Since the etiologic factor in normal grief is a distinct sustaining cause, normal grief does not qualify as a malady according to our definition, and we do not believe it should be so regarded. This conclusion may seem to be inconsistent with our claim that someone has a malady if his condition persists for any significant period of time after it has been caused by some distinct factor. If someone has been deprived of food for several days, we say he has a malady if his condition does not improve quickly after eating. If he were to be weak for days or weeks before resumed eating made him feel better, we would say that he suffered from malnutrition, which is a malady. Since grief lasts for weeks, if not months, why doesn't it count as a malady even though its cause is clearly distinct from the person? Our answer is that the loss continues; when one grieves for the loss of a loved one, the loss remains for the whole period of the grief and beyond. It is not that after the first day there is no longer any loss; that is, it is not like the condition of the person who has not eaten for several days and then is given a normal meal. It is more like that of someone who is continually given less food than is necessary to sustain normal functioning.

A reasonably close physical analogy to grief is the one- to two-month period of distress which low landers experience if they move to a high mountainous location. Their distress is due to the lower concentration of oxygen in the air and will gradually disappear as their hematologic and respiratory systems adjust. The lower concentration of oxygen is a distinct sustaining cause; if the people returned to sea level, their distress would very quickly cease. It would usually be said of such persons that they were going through a "normal adjustment reaction" but not that they had a malady.

The evils suffered by both the griever and the new mountain dweller may be nearly identical to the evils that would be suffered if either person did have a malady, that is, if there were no distinct sustaining cause for the evils. Nonetheless, in the case of the griever or new mountain dweller, the suffering of the evils is explained as due to something abnormal in the person's environment, and not to something wrong with the person himself. For conceptual precision, it is useful to reserve "malady" for conditions in which the suffering is explained primarily as due to something wrong with the person rather than something abnormal about his environment.

It is because the evils suffered in maladies and nonmaladies can be so similar (e.g., grief and an endogenous depressive episode) that if, like Engel, one does not have the concept of a distinct sustaining cause it is so easy to conflate the two conditions. This conflation is evident at a later point in Engel's paper. He imagines a "skeptic" raising "some pointed questions:"

> Is not grief simply a natural reaction to a life experience? How can one put it in the same category as the pathological states we call disease? (p. 19)

Engel replies:

> To this we answer that it is "natural" or "normal" in the same sense that a wound or a burn are the natural or normal responses to physical trauma. The designation "pathological" refers to the changed state and not to the fact of the response. That one responds to thermal radiation with a burn is natural or normal. The burn itself constitutes a pathological state and the concept is as appropriately applied to the state of grief as to a wound, burn, or infection. (p. 19)

But wounds and burns are not the same as grief; they have no distinct sustaining causes. The thermal radiation causes the burn, but after the thermal radiation is removed the burn remains. In fact, it is only if pain and suffering remain that one says a burn has occurred at all. Engel is correct that both grief and burns represent responses that are normal (if that means statistically common in the presence of certain stimuli) and pathological (if that means associated with the suffering of evils), but showing that conditions are alike in some respects does not mean that they are alike in others (e.g., in their continued dependence on the environment).

In the case of grief and of moving to a high altitude, the loss of the loved one and the lower amount of available oxygen both continue to function as distinct sustaining causes. They are not like the thermal radiation which completely disappears, leaving behind only a changed condition of the person. Even though the distinct sustaining causes persist, however, the evils normally attenuate over a period of time. If the evils persist much longer than is normal for the species, for example, if the grief period is significantly prolonged, then one does begin to explain the suffering as due to something wrong with the person, and it is intuitively plausible to say that prolonged or extreme grieving is a malady, just as prolonged or extreme failure to adjust to high altitude would count as a malady.

It is interesting to note that DSM-III (1980, p. 333) does not classify "uncomplicated bereavement" as a mental disorder; it is instead given a V code (see below for an explanation of V codes). We agree with DSM-III's classification, but we fail to see why normal grief does not satisfy DSM-III's definition of a mental disorder. It is (1) a clinically significant behavioral and psychological syndrome, (2) it is associated with both painful symptoms and impairment in functioning, (3) the inference could be made that there is psychological dysfunction, and (4) the disturbance is not in the relationship between the individual and society. The authors could counter by saying that in normal grief there is no psychological dysfunction. But they acknowledge that "a full depressive syndrome" frequently develops, including depression, anorexia, weight loss, and insomnia. Persons with such symptoms would ordinarily be regarded as having some type of psychological dysfunction, according to the admittedly vague connotations of that term. In this case, we believe that the authors of DSM-III correctly followed their intuitions and not their definition.

Mental maladies

At the beginning of Chapter 4, we suggested that the most straightforward way of answering the question of whether the conditions treated by psychiatrists qualified as maladies was to investigate the extent to which these conditions satisfied the criteria of an adequate definition of malady. We are now in a position to carry out this task and want to apply our definition to a variety of mental conditions, some that clearly qualify as maladies and others that do not.

Schizophrenia, as defined in DSM-III (1980, pp. 188–191), clearly qualifies as a malady. Schizophrenic individuals suffer from a variety of symptoms, including bizarre delusions, persecutory delusions, incoherence, and disorganized behavior. These symptoms result in varying degrees of impairment in interpersonal and vocational functioning. Most of the symptoms of schizophrenia are psychologically painful to experience and thus directly involve the suffering of an evil. Schizophrenics also often have irrational desires as well as irrational beliefs (delusions); as explained in Chapter 2, both increase the risk that an affected individual will suffer evils in the future.

The evils suffered (or likely to be suffered) by schizophrenics do not have distinct sustaining causes. Even if one favors an etiological theory emphasizing early life experiences, the most that one could claim is that environmental factors originally caused the condition, not that they currently sustain it. The effects of the early experiences have been assimilated and integrated so thoroughly into the individual that there is no plausibility in regarding them as distinct sustaining causes.

Bipolar disorders and major depressive episodes clearly qualify as maladies as well. The evils suffered during major depressive episodes are obvious: sadness; loss of energy, indecisiveness, feelings of worthlessness, and so on (see DSM-III, 1980, pp. 213–215). The evils which the manic patient suffers, or is at significantly increased risk of suffering, can of course be just as intense despite the patient's partly euphoric mood: grandiose delusions, reckless driving, squandering of money, and so on (see DSM-III, 1980, pp. 208–210).

Phobias and compulsions of any significant degree qualify as maladies. Phobic patients suffer from the dread of the feared object and frequently suffer from a volitional disability and a consequent disruption of their lives occasioned by the necessity of avoiding the feared object; they often suffer from panic attacks as well (see DSM-III, 1980, pp. 225–232).

Compulsive patients also frequently have a volitional disability, and it may also disrupt their lives: they are unable to will not to carry out their compulsive activity for a period of time. For example, a compulsive hand washer may be unable to refrain from washing his hands less than two or three times an hour. And some compulsions, like hand washing, directly cause other evils as well, such as sore, excoriated skin (see DSM-III, 1980, pp. 234–235). Thus, both phobias and compulsions are characterized by the suffering of various evils, and in neither condition is a distinct sustaining cause present. These disorders are discussed in more detail in the next chapter.

The intensity of the evils suffered determines the intensity of the malady. The analogous point was made with respect to physical maladies in Chapter 4 and applies in just the same way here. An agorophobic woman who is unable to leave her house for months, during which time she suffers from nearly daily panic attacks, has a very severe malady. A bachelor who must leave his office once or twice daily and drive the three miles to his home to be sure it is not on fire has a somewhat less severe malady, though it may very well qualify as an Obsessive Compulsive Disorder in DSM-III (1980), which requires that the compulsion be "a significant source of distress to the individual or interfere with social or role functioning" (p. 235). And some maladies may be relatively minor and comparable, say, to a mild allergy. For example, a woman might be chronically troubled with mild anxiety symptoms which were insufficient in number and/or intensity to meet the DSM-III (1980) criteria for Generalized Anxiety Disorder (pp. 232–233). While it would be true that she was suffering evils in the absence of a distinct sustaining cause, her condition would represent only a minor malady.

Mental conditions that are not maladies

There are mental conditions in which one suffers evils for which there is an identifiable distinct sustaining cause; these do not qualify as mental maladies according to our definition. For example, a man who has functioned happily as an employee over the years may dislike his new unpleasant supervisor and often suffer tension and anger when they are together. Other workers may feel the same way about this supervisor. The employee would not have a malady because the supervisor serves as a distinct sustaining cause for the evils the man is experiencing. If the supervisor were to be replaced, then the evils would presumably end as well. The details of this case can be changed so that the man would have a malady: suppose that he has held a long succession of jobs and in all of them has resented his supervisors and suffered tension and anger in their presence. It would then seem likely that there was something wrong with the man himself, for example, in his relationships with those having authority over him. If the evils he suffered were at all significant, then he would have a mental malady.

DSM-III (1980) has provided a special category called "V Codes for Conditions not Attributable to a Mental Disorder that are a Focus of Attention or Treatment" (pp. 331–334). Many of the specific V codes listed refer to conditions in which persons are suffering evils but in which, according to our analysis, a distinct sustaining cause is present. An example is "V62.20 Occupational Problem," which is described as a category to be used "when a focus of attention or treatment is an occupational problem that is apparently not due to a mental disorder" (p. 332). The first man described above might be given this code were he to come to a mental health center for help in deciding whether to change his job. Other V codes in which the suffering of evils might be due to a distinct sustaining cause are "V62.30 Academic Problem," "V61.20 Parent–child Problem," and "V61.10 Marital Problem." Of course, in each of these problem areas, individuals often suffer evils that are due not only to distinct sustaining causes but also to something wrong with themselves, and therefore V codes could not be appropriately used.

While some of DSM-III's V codes refer to conditions in which evils are suffered because of a distinct sustaining cause, other kinds of conditions are given V codes as well. In one interesting group of conditions, the behavior that defines the condition is intentional, voluntary, and rational, such as "V65.20 Malingering" and "V17.01 Adult Antisocial Behavior" (an example of the latter being a professional thief). We agree with DSM-III that these conditions are not maladies. In the next chapter, we analyze intentional and voluntary behaviors and relate our analysis to many DSM-III categories. But first, we want to examine DSM-III's account of "Psychosexual Disorders," which seems to be the least satisfactory section of the manual.

Sexual Maladies

We will begin by sketching our own view of what kinds of sexual conditions qualify as maladies and then discuss the DSM-III category of Psychosexual Disorders. According to our account, someone has a sexual malady if and only if his sexual condition, other than his rational beliefs and desires, is associated with the suffering or increased risk of suffering evils in the absence of a distinct sustaining cause.

It is useful to make several distinctions.

1. *Behavior versus fantasy*. Some sexual conditions involve engaging in a particular kind of sexual behavior: having homosexual relations, having heterosexual relations, exhibiting oneself, masturbating while caressing a fetish object, and so on. Other sexual conditions only involve recurrently imagining certain behaviors and situations, and often this is done with little or no increase in the likelihood that the person will ever actually engage in the behavior. For example, a person might imagine having sexual contact with an animal, and even experience some sexual excitement during the fantasy, and yet have no inclination to ever actually engage in the behavior.

There are intermediate cases in which persons have recurrent fantasies together with desires to actually engage in the imagined behavior, though these desires are not acted upon. For purposes of simplicity, we will divide these intermediate cases into two categories: those that (significantly) increase the probability of engaging in the behavior and those that do not. We shall treat the latter as pure fantasy and the former as if it were behavior. This oversimplifies the matter, but it should not prove misleading. It is important to note that except in very unusual cases, it is only recurrent fantasies and behaviors that are considered conditions of the person. An occasional act or fantasy would be just that, an act or fantasy, and not a (continuing) condition of the person.

2. *Ego-syntonic and ego-dystonic fantasies*. It is also useful to distinguish between sexual fantasies which are distressful and unpleasant for the person to experience, that is, are ego-dystonic, and those that are not, that is, are ego-syntonic. Note that neither ego-dystonic nor ego-syntonic fantasies need be acted upon. For example, a person may acknowledge having some pedophilia fantasies, feel no particular distress about having them, and yet have no desire that increases his probability of acting upon these fantasies.

According to our account, *ego-syntonic fantasies* alone are never maladies because no evil or risk of evil is being suffered. This is true no matter how unusual or bizarre the fantasy is; if it causes no distress to the person, and if it does not increase the probability that at some later time evils will be suffered, it is not a malady.

Recurrent *ego-dystonic fantasies* are maladies. Sometimes the malady may be very minor, for example, a person who occasionally fantasizes engaging in homosexual behavior, feels a moderate degree of homosexual excitement, and consequently feels some distress. In other cases, the malady may be quite serious, for example, a man who is intensely distressed by his constantly recurring fantasies about sexual behavior with dead persons. Both examples, qualify as maladies according to our definition in that the persons are suffering evils in the absence of a distinct sustaining cause.

3. *Ego-syntonic and ego-dystonic behaviors.* First, we will discuss ego-syntonic sexual behaviors. Such behaviors are usually not maladies because the person is suffering no evils. Unlike sexual fantasies, however, some ego-syntonic behaviors can be maladies nonetheless. This can occur in either of two ways. The first way ego-syntonic sexual behavior can represent suffering an evil, though it is not distressful, occurs when the sexual behavior is not under voluntary control, that is, when the person has a volitional disability. For example, a man may feel no personal distress about his desire to dress in women's clothes and walk around the city in which he lives. But suppose he is presented with numerous and varied noncoercive and coercive incentives for not engaging in this behavior: his wife, whom he loves, threatens to leave him; his job, which means very much to him, is jeopardized by his behavior; he is part of a community which he respects and which will ostracize him. If he persists in frequently cross-dressing, then it is likely that he suffers from a volitional disability with regard to cross-dressing.

This example may seem bizarre, for it would seem that with all of these negative incentives his cross-dressing behavior could no longer be ego-syntonic. But when we call the behavior ego-syntonic, we mean only that he is not distressed by the behavior itself; we do not imply that he is not distressed by the consequences that are contingent on that behavior. Granted that if he tried to stop and couldn't, the behavior itself might come to be distressing, but even before that happens he is suffering the evil of a volitional disability. Someone in jail might not be distressed by his lack of freedom but would be suffering the evil of deprivation of freedom just the same. Thus some ego-syntonic sexual behavior may reflect a volitional disability.

A second way in which an ego-syntonic sexual behavior might qualify as a malady is if it has a high likelihood of being generally known to or discovered by others, and the nearly universal reaction of others would be one of repugnance and revulsion. Such recurrent sexual behavior would show that the person has a malady because his sexual behavior would lead to a degree of deprivation of freedom associated with being the target of general repugnance. Such sexual behaviors as necrophilia or having sex with very young children are examples. These sexual conditions would thus be analogous to the physical condition of a person with a serious physical disfigurement who was not distressed by it, discussed in Chapter 4. The vast majority of unusual sexual behaviors, of course, are not like necrophilia or having sex with very young children; that is, they can be carried out with a high likelihood that others will not know (e.g., cross-dressing in private, or any behavior between two consenting adults) and / or they are not met with universal revulsion (e.g., homosexuality).

We have never encountered anyone whose sexual behavior is universally considered to be repulsive, but who nonetheless claims it is ego-syntonic. The male patients we have known who have had sexual relations with very young children have been very distressed about their behavior and thus have had a malady on that basis alone. We have never known a necrophiliac but would guess that this behavior is also likely to be quite ego-dystonic.

The preceding discussion makes it clear that the primary reason why certain recurring sexual behaviors are maladies is that they are ego-dystonic. The person engaging in the behavior is distressed by it. Of course, such behavior is probably also a manifestation of a volitional disability, but even if it is not, the distress, if significant, is sufficient to make it count as a malady. Note that neither in the case of distress nor of a volitional disability is the sexual condition a malady because it is sexual, but rather because of some other characteristic attached to the condition. Thus, we believe that when homosexuality qualifies as a malady it is because of the distress the person experiences, not because of the person's homosexual fantasies or desires. Similarly, as noted above, cross-dressing is not a malady but involuntary cross-dressing is because the person suffers from a volitional disability. The only reason for listing the various paraphilias as sexual maladies is that a high percentage of persons with these fantasies or behaviors apparently suffer distress or volitional disabilities. We suspect that it is the belief that there is not a similar high percentage of persons who masturbate who suffer distress or a volitional disability that accounts for the lack of a category of ego-dystonic masturbation in DSM-III.

4. *Paraphilias and homosexuality in DSM-III.* "Paraphilia" is the general term used in DSM-III (1980, pp. 266–275) to describe eight kinds of sexual conditions: fetishism; transvestism; zoophilia (sexual excitement via animals, either in fantasy or reality); pedophilia (sexual excitement via

children, either in fantasy or reality); exhibitionism; voyeurism; sexual masochism; and sexual sadism. In addition, there is a residual category, "atypical paraphilia," for such conditions as necrophilia (sexual excitement via dead bodies, either in fantasy or reality) and "telephone scatologia" (sexual excitement via the making of lewd phone calls). DSM-II (1968) called these kinds of conditions "sexual deviations" and included homosexuality in the group. DSM-III lists homosexuality separately from the paraphilias (pp. 281–282) and only considers ego-dystonic homosexuality to be a mental disorder (later printings of DSM-II made a similar change).

There are serious difficulties with DSM-III's analysis of the sexual maladies. Many individuals who engage in the kinds of "paraphiliac" fantasies and/or behaviors described in DSM-III do not have a malady (according to our definition) and may not even have a mental disorder according to DSM-III's own definition. Their sexual condition may not be ego-dystonic, and often no evils or increased risk of evils are being suffered. In DSM-III terminology, there is often no distress or impairment of functioning. We suspect that the reasons for DSM-III's labeling these conditions as mental disorders, whether or not they are ego-dystonic for the individual or involve other evils, are primarily historical and dependent on certain psychological theories (see below). Even in the case of homosexuality, where the criterion of whether the condition is ego-dystonic plays an important role, the description of the malady is marred by adding that these individuals must also be distressed by the fact that they have weak or absent heterosexual arousal (pp. 281–282). Thus, a bisexual person, who has strong heterosexual arousal, but who is severely distressed by and comes for treatment of his homosexual impulses and behavior, cannot be diagnosed as having ego-dystonic homosexuality.

It seems that the authors of this section of DSM-III have adopted the psychological theory that heterosexuality (fantasy and behavior) between consenting adults is the ideal and that except for ego-syntonic homosexuality the absence of heterosexuality is ipso facto a mental disorder. The section on paraphilia is in fact introduced with the sentence, "The essential feature of disorders in this subclass is that unusual or bizarre imagery or acts are necessary for sexual excitement" (p. 266). Thus, it is deviance that defines disorder, not, as in our account and as implied by the DSM-III definition of mental disorder, the suffering of evils. In fact, it is noted on the next page of DSM-III, "Individuals with these disorders tend not to regard themselves as ill, and usually come to the attention of mental health professionals only when their behavior has brought them into conflict with society" (p. 267). But this seems to conflict with DSM-III's own definition of mental disorder, which requires that the individual typically suffer either distress or impairment in functioning, and that his "disturbance" not be limited to a conflict with society. The only way the ego-syntonic paraphilias can be classified as mental disorders (using DSM-III's definition) is to believe that they involve an important impairment in functioning. This can be done only by assuming that those whose sexual fantasies or behavior do not involve consenting adults (preferably of the opposite sex) are thereby suffering an important impairment in functioning.

DSM-III's treatment of several of the other paraphilias, such as fetishism, zoophilia, pedophilia, suffers a similar problem: whether the condition is ego-dystonic is ignored, and the defining criteria for the disorder are whether some deviant fantasies and/or behaviors are present and whether heterosexual activity is weak or absent. The manual is inconsistent concerning how much normal heterosexuality is allowed. The introductory sentence (quoted above) stipulates that "unusual or bizarre imagery or acts are necessary for sexual excitement" (p. 266). In the specific criteria for several of the conditions (fetishism, zoophilia, pedophilia, and voyeurism), however, it is often stated that the deviant imagery or acts are the "repeatedly preferred or exclusive method of achieving sexual excitement." If something is only "repeatedly preferred," it is not necessary. We suspect the authors fudged a little because they were quite aware that making peeping a necessary condition for sexual excitement would eliminate a great deal of voyeurism. Conditions like voyeurism exist along a continuum. A few men can attain sexual excitement only while peeping; others may prefer peeping but also engage in occasional or frequent heterosexual relations; and still other actively heterosexual men engage in voyeurism only rarely. On our account, they all have a sexual malady to the extent that their peeping is ego-dystonic, but we have seen how DSM-III's definition of mental disorder allows only the exclusive peepers to be so classified.

It is in the criteria for transvestism (pp. 269–270) that the use of deviance rather than suffering evils in defining a mental disorder creates the most obvious problems for DSM-III. It is well known that a large number of transvestite men have quite active heterosexual sex lives, so it would have been very difficult for the authors to have included the criterion "repeatedly preferred or exclusive method of achieving sexual excitement" here. Since cross-dressing is certainly deviant, however, a very odd criterion is now included: "Intense frustration when the cross-dressing is interfered with"

(p. 270). At the very least, this criterion seems quite ad hoc; it is included in none of the other paraphilias. But more important, it does not seem to describe a unique characteristic of transvestites; most people are apt to be intensely frustrated when their preferred style of sexual activity is interfered with. Indeed, many people experience intense frustration whenever any strong desire is repeatedly interfered with. Perhaps this criterion was included in order to impart a measure of urgent distress to the condition of transvestites so that they could at least seem to satisfy DSM-III's definition of mental disorder.

What is at stake here is much more than verbal quibbling. It is important that criteria for specific mental maladies, like those for specific physical maladies, always involve the suffering or increased risk of suffering evils. Otherwise, people who do not suffer evils will be said to have maladies just because they arc deviant, and less deviant people who do suffer will not count as having a malady.

For example, if a man with an active heterosexual sex life is deeply troubled because on occasion he cannot resist the impulse to fondle young girls sexually, he may not satisfy DSM-III's criteria for pedophilia (pp. 271–272) because children may well not be his repeatedly preferred or exclusive outlet. Yet it would be counterintuitive and unfortunate not to say that he has a malady, in that he is suffering a good deal from a sexual condition. Whether he fondles young girls frequently or infrequently seems less important.

We realize that DSM-III does not explicitly state that sexual deviance is a sufficient condition for a mental disorder but we believe that, in fact, it lists some of the paraphilias as mental disorders primarily for this reason. Unusual sexual fantasies and behaviors may tend to be ego-dystonic and thus maladies but, if true, it is still the suffering of evils and not the deviance which is essential in labeling these conditions maladies. As we have seen, unusual sexual fantasies and behaviors can also be ego-syntonic and not be maladies. Also, a common sexual behavior like masturbation can be very ego-dystonic and thus qualify as a malady.

Thus on our account, ego-syntonic sexual deviance in fantasy or behavior in and of itself is no more a malady than is ego-syntonic political deviance. The tendency to regard deviance as a sufficient feature of sexual disorders leaves psychiatry open to the justified criticism that deviance per se, sexual or political, has become closely linked in psychiatric thought with sickness. It is not surprising that many individuals fear the possible repercussions of that kind of thinking.

Notes

1. See Roth (1976) for further examples of physical maladies of unknown pathophysiology.
2. Similar arguments have been developed by Moore (1975), Margolis (1976), Roth (1976), and Farrell (1979).
3. For a fuller explanation of the concept of voluntary abilities, see Duggan and Gert (1967).

References

DSM-II. Diagnostic and Statistical Manual of Mental Disorders. Washington, D.C.: American Psychiatric Association, 1968.

DSM-III. Diagnostic and Statistical Manual of Mental Disorders. Washington, D.C.: American Psychiatric Association, 1980.

Duggan, Timothy, and Gert, Bernard. Voluntary abilities. *American Philosophical Quarterly,* 1967, 127–135.

Engel, George. Is grief a disease? *Psychosomatic Medicine,* 1961, *23*, 18–22.

Farrell, B. A. Mental illness: A conceptual analysis. *Psychological Medicine,* 1979, *9,* 21–35.

Margolis, Joseph. The concept of disease. *The Journal of Medicine and Philosophy,* 1976, *1,* 238–255.

Merck Manual of Diagnosis and Therapy, 13th ed. Rahway, N.J.: Merck Sharp & Dohme, 1977.

Moore, Michael S. Some myths about mental illness. *Archives of General Psychiatry,* 1975, *32,* 1483–1497.

Roth, Martin. Schizophrenia and the theories of Thomas Szasz. *British Journal of Psychiatry,* 1976, *129*, 317–326.

6

Volitional Disabilities

As noted in the last chapter, one of the most common evils associated with mental maladies is a particular type of disability, namely, a volitional disability. We believe that the concept of a volitional disability is a very useful one for medicine and psychiatry. Everyone recognizes that phobias and compulsions are similar in a very important respect, but until now there has been no concept that makes this explicit. Our account of a volitional disability not only makes explicit the similarity between phobias and compulsions, it also brings out the similarity between alcoholism, some kinds of noncompliant patient behaviors, and compulsions and phobias. In addition, it allows us to provide a useful classification schema for human actions.

Voluntary abilities[1]

In the previous chapter, we referred briefly to voluntary abilities and noted that we do not usually distinguish between the components of a voluntary ability unless somcone is disabled in some way. We realize that normally we do not encounter physical, cognitive, or volitional abilities independently of each other, but we do test for abilities of each of these three types, and so it is worthwhile to provide an analysis of each. Having such an analysis allows us to see both the similarities and differences among the three types of ability.

Physical and cognitive abilities have an identical analysis. They differ from each other only in the way indicated in the last chapter; physical abilities depend on some particular part or parts of the body other than the brain, and cognitive abilities do not. For both kinds of abilities, the following analysis seems to capture what is essential.

A has the physical (cognitive) ability to do X (a kind of action) if, and only if, for a reasonable number of times, if A were to will to do a particular act of kind X, then given a reasonable opportunity, A would do that particular act of kind X.

Examples show clearly the point of the analysis. If we want to know if someone has the physical ability to touch his toes or the cognitive ability to memorize ten lines of poetry, we provide situations in which he wills to demonstrate his ability and has a reasonable opportunity to do so. If he does exercise that ability a reasonable number of times (it is not necessary that he touch his toes every time he tries, or never fails to recite the ten lines correctly), then we say he has the ability in question.

Sometimes we know that someone has a physical (cognitive) ability even though we have never seen him demonstrate it, because he has demonstrated other more complex physical (cognitive) abilities. And since physical (cognitive) abilities do not depend on context, if someone demonstrates a physical ability in one context but not in another, we are justified in concluding that it is not the physical ability that comes and goes, but that in one context he does not will to exercise that ability. However, if someone never exercises a physical (cognitive) ability in a certain context, even when there are very strong incentives for doing so, we may come to think that he lacks the volitional ability to exercise that physical ability in this context. For example, if someone never exercises his physical ability to walk when this would result in his entering an elevator, then we do not conclude he lacks that physical ability: rather, we may conclude that he lacks the volitional ability to will to enter an elevator.

Volitional ability[2]

A volitional ability, like all other abilities, is always related to a *kind* of action. When we say that A has the volitional ability to will to do X, X stands for a kind of action, like entering elevators. Or the kind of action may be described in more general terms, such as entering small enclosed spaces. Since we are primarily interested in those volitional abilities that people lack, we usually describe the kind of action as the most general kind that one lacks the volitional ability to will to do. This is not merely so that we can say a person has claustrophobia, rather than several similar phobias, such as elevator phobia or darkroom phobia. Rather, it is the most general feature that is psychologically significant and hence most accurately describes what one lacks the volitional ability to will to do.

If someone has a severe phobia about being in elevators, he may not have the volitional ability to will to enter an elevator. (We say "may not" rather than "does not" because phobias do not always involve a volitional disability; they sometimes involve only the suffering of inappropriate anxiety. This will be discussed later in the chapter.) Someone with an elevator phobia usually has the requisite physical abilities but does not have the volitional ability to will to exercise them when confronted with an open elevator door. Our discussion here will focus on whether a person does or does not have the volitional ability to will to carry out some kind of action, and we will assume that the person does have the required physical or cognitive abilities to carry out the action.

Our analysis of a volitional ability is as follows: S has the volitional ability to will to do X, if and only if:

1. if S believes that there are coercive incentives for doing a particular act of kind X; he would almost always will to do it.
2. if S believes that there are noncoercive incentives for doing a particular act of kind X; he would, at least sometimes, will to do it.
3. if S believes that there are coercive incentives for *not* doing a particular act of kind X; he would, almost always, will *not* to do it.
4. if S believes that there are noncoercive incentives for *not* doing a particular act of kind X; he would, at least sometimes, will *not* to do it.

A coercive incentive is one which it would be unreasonable to expect *any* rational person not to act on and hence provides an excuse for so acting. A noncoercive incentive is an incentive which is not coercive. There are many different kinds of incentive: moral, prudential, patriotic, and so on. Moreover, while money can serve as an incentive, one sum of money, such as $100, is obviously a weaker incentive than a larger sum of money, such as $5,000. However, in normal circumstances, no amount of money is a coercive incentive. Coercive incentives must involve suffering significant evils, such as death and serious disability, for only these can provide the kind of incentive that it would be unreasonable to expect any rational person not to act on. (To be subject to coercive incentives does not mean one is a victim of coercion. Coercion always involves an intentional threat by another person, but coercive incentives can come from anywhere.) However, even significant evils are not always coercive incentives. Whether an incentive is coercive or noncoercive depends not only on the strength of the incentive but also on the act for which it provides the incentive. That one will lose one's arm by not undergoing the slight pain of an injection provides a coercive incentive for having the injection. However, the same incentive, avoiding losing one's arm, is not a coercive incentive for undergoing multiple painful operations for several years.

An example of someone who lacks a volitional ability would be a compulsive hand washer who believes that there are both coercive and noncoercive incentives for not washing his hands once every waking hour and yet does not act in accordance with these incentives, and indeed laments his failure to act. Such a person lacks the volitional ability to will to wash his hands once every waking hour, even though he does intentionally wash his hands. Compulsive action of this kind provides the clearest example of someone who acts intentionally yet not voluntarily. As we explain more fully later in this chapter, we can account for this category of action by noting that having a volitional ability to will to do X requires willing to refrain from doing particular acts of kind X when there are appropriate incentives for so willing.

Though we talk of willing and the volitional ability to will, we are not committed to some special faculty of the will or to some special internal act of willing. For us, one wills to do X if and only if one intentionally does or tries to do X. We offer no philosophical analysis of willing, noting only that as we use the term there is no temporal gap between willing and doing. But, as noted above, willing to do X does not imply having the volitional ability to will to do X, just as shaking by a palsy victim does not imply the physical ability to shake. To have the ability one must be able both to will (do) and refrain from willing (doing). This is why one can do something intentionally (will to do it) and yet not do it voluntarily (not have the volitional ability to will to do it).

We do not make any claims about what is responsible for the lack of a particular volitional ability. What we say is compatible with the claims of both psychoanalysis and behaviorism. Concerning the compulsive hand washer, we can say that guilt feelings due to real or imagined doing of some forbidden act, such as masturbation, create such anxiety that he does not have the volitional ability to will to refrain from washing his hands. Or we can say that he was conditioned to behave according to a particular schedule of reinforcement so that he cannot act in any other way.

The ability to believe

We have characterized volitional ability in a hypothetical manner which seems similar to the hypothetical manner in which physical and cognitive abilities were analyzed. However, whereas one does not lack one of these latter abilities if he never wills to exercise it, one may lack the volitional ability to will to do X if he never believes in the existence of coercive and noncoercive incentives for doing X. Someone may lack the ability to believe in the existence of incentives for certain kinds of actions. Thus, though he might have the volitional ability if he believed, he lacks the ability to believe. The volitional ability to will to do X (a kind of action) presupposes the ability to believe that there are incentives for and against doing X. Thus a complete understanding of volitional ability requires understanding the ability to believe.

We propose the following analysis of the ability to believe: S has the ability to believe some proposition (P), if and only if, if S were presented with what almost everyone with similar knowledge and intelligence would regard as overwhelming evidence that P were true, then S would believe that P was true. And if S were presented with what almost everyone with similar knowledge and intelligence would regard as overwhelming evidence that P were not true, then S would believe that P was not true.

It is important to note that this analysis of the ability to believe is not limited to beliefs that there are coercive and noncoercive incentives for both doing and not doing X. The ability to believe or the lack of that ability applies to all kinds of beliefs, for example, that one has cancer, that one's wife is faithful, that one's son has been killed. It is interesting to note the strong connection between holding an irrational belief (see Chapter 2) and lacking the ability to believe. We suggest that irrational beliefs are symptoms of mental maladies because they count as the lack of the ability to believe. What is wrong with having psychotic delusions, such as paranoid delusions, is not that they are false but rather that they show the person does not have the ability to believe; he does not respond to the overwhelming evidence in a way that is appropriate for someone with his knowledge and intelligence. This explains in a more precise way why holding irrational beliefs is a symptom of a mental malady.

A classification of actions

In this section, we provide a general framework for classifying human behavior by focusing on the concept of excuses. Excuses are usually invoked in order to lessen or eliminate responsibility for doing something wrong or for the bad consequences of something one has done. Generally speaking, in offering excuses, we try to show that our relationship to the bad consequences for which we are being held responsible differs in a significant way from a case in which we must accept full responsibility, that is, one in which (1) we acted intentionally in order to bring about those bad consequences; (2) we did not, at the time, suffer from any relevant volitional disability, and (3) our intentional action was not due to our being subject to coercive incentives.

In other words, the paradigm case of being held fully responsible for bringing about bad consequences is that of a person who acted intentionally, voluntarily, and freely. We call such actions *free* actions (A). Thus in offering excuses, one move is to try to show that what was done was not done freely, that is, as the result of coercive incentives. We call such actions *unfree* actions (B). Another move is to show that the action was not done voluntarily, that is, it was due to some volitional disability. We call such actions *unvoluntary* actions (C). A third move is to deny that we intended to bring about the consequences for which we are being held responsible. We call such actions *nonintentional* (D). A fourth move is to claim that though the consequence stems from some movement (or lack of movement) of our body, that movement is not properly described as an action of ours at all. We call such movements (or lack of movements) *nonactions* (E). The following diagram shows how these various categories are related to each other.

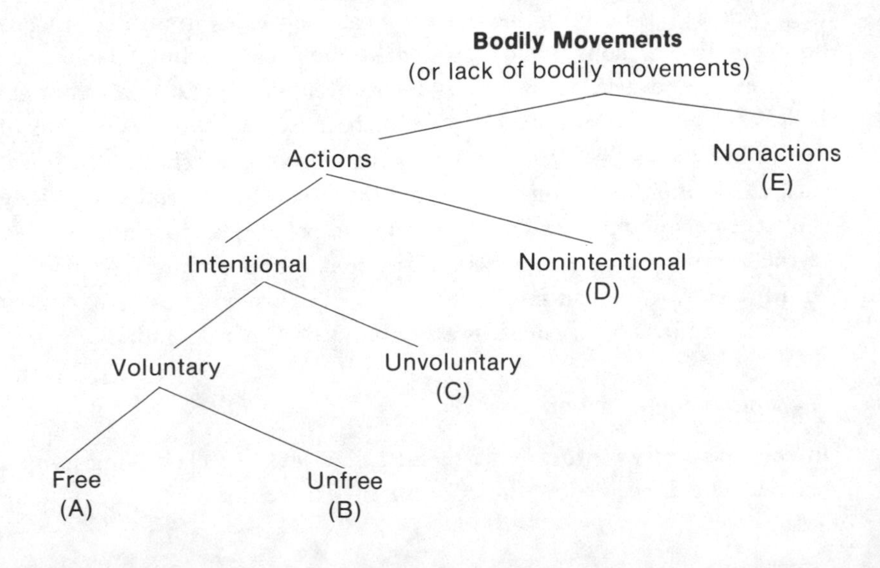

We do not claim that it is always possible to categorize a given movement clearly. But often no distinction is made between category E and category D; both are simply lumped together as nonintentional actions. However, though there are significant borderline cases, there are also clear examples of each category. If A is pushed by B against C, then when C complains, A can truthfully say, "I didn't do anything; I was pushed by B." This kind of case is the clearest example of nonaction. But there are other cases which also seem to be nonactions, such as movements of a person in an epileptic seizure, reflexive behavior such as knee jerks and eye blinks, and all the movements of a newborn infant. Thus what some call involuntary actions, we regard as non-actions. Less clear cases would be complex movements made during sleep, or movements made by someone experiencing a "Sleepwalking Disorder" (DSM-III, 1980, pp. 82–84). We think that providing a distinct category of nonactions allows us to use the term "action" in a philosophically more fruitful fashion.

The clearest examples of actions in category D are accidents (e.g., in reaching for the salt, I knock over my glass of water) and mistakes (e.g., I put a teaspoon of salt into my tea instead of sugar). All of the clear cases of actions that fit into category D involve a person who is intentionally doing something but not intentionally doing the action (bringing about the consequences) at issue: for example, a hunter who, while shooting at a bird, wounds the picknicker behind the bush. Category D, therefore, unlike category E, typically involves some intentional action on the part of the agent.

Before we discuss category C, unvoluntary action, some discussion of the two categories of voluntary action, that is, A, free action, and B, unfree action, is in order. All voluntary action is intentional; that is, it is willed. What makes it voluntary is that it is willed by someone with the volitional ability to will it. An act done freely, or a free action (A), is thus one that is both intentional and voluntary; that is, it is willed by a person who has the volitional ability to will to do that kind of action, and what makes it free is the fact that no coercive incentives have been responsible for the occurrence of the action. Normally, an unfree action is also one that is willed by someone with the ability to will to do that kind of action. It is unfree because there are coercive incentives which are responsible for the action's occurrence. A classical example of such an unfree though voluntary act is Aristotle's case of the sea captain who, in order to save his ship and crew, jettisons his cargo in a storm.

Some writers have combined categories B and C and regarded both as examples of actions done because of compulsion. We believe, however, that in order to avoid confusion, it is important to distinguish between unfree actions (B) and unvoluntary ones (C). We regard unfree actions as those that are caused by something distinct from the agent, namely, coercive incentives. Were these coercive incentives removed, the person would be free to refrain from doing the contemplated action. However, we regard unvoluntary actions as those caused by something about the agent himself, namely, a specific volitional disability. We hold that in no normal situation would such a person act differently than he did in fact act. The typical use of the term "compulsion" to cover both categories B and C is unfortunate, for it leads one to regard "inner compulsion" as identical to external compulsion, such as coercion, in all respects except that it comes from within the person rather than from without. But this obscures the fact that external compulsion, such as coercion, typically affects the doing of particular actions—your money (here and now) or your life—whereas inner compulsion, that is, a volitional disability, always affects doing or refraining from doing kinds of actions.

It is category C which we regard as most important. We have already seen that there is a tendency to equate categories C and B; however, there is an even stronger tendency to equate categories C and D. "Voluntary" and "intentional" are sometimes used as synonyms, and thus it is natural to equate unvoluntary with nonintentional action. But the introduction of the concept of volitional ability shows clearly that there is an important distinction between voluntary and intentional action. An intentional action is one that is willed. A voluntary action is one that is willed by someone with the volitional ability to will to do that action. The standard confusion has arisen because it was assumed that if one willed to do X, then one must have the volitional ability to will to do X. Consideration of the compulsive hand washer shows the falsity of this assumption. The compulsive hand washer intentionally washes his hands; he deliberately goes to the sink, takes the soap and lathers up. He knows what he is doing and wills to do it. However, he does not do so voluntarily. He lacks the volitional ability to will to wash his hands. This paradoxical situation, of willing what one does not have the volitional ability to will, arises because having the volitional ability to will to do X includes as a necessary feature willing not to do X in appropriate circumstances. The volitional ability to will to do X requires that one can will not to do X. A compulsive hand washer wills to wash his hands, but since he cannot will not to wash his hands, he lacks the volitional ability to will to wash his hands. Similarly, the claustrophobic wills not to enter the elevator but he does not refrain voluntarily, for he cannot will to enter the elevator, thus showing that he lacks the relevant volitional ability.

It is the concept of volitional ability which allows us to distinguish category C from categories B and D and to make it a category in its own right. This is especially important in applying this classification scheme to psychiatry, for some significant psychiatric disorders involve category C.

Applications to psychiatry

It is instructive to apply the above classification of actions to various conditions seen by psychiatrists. The symptoms of a great many mental maladies are characterized, at least in part, by some type of unvoluntary behavior, but other categories of actions are also relevant. We will give examples of conditions in four categories: intentional voluntary actions (category A); intentional unvoluntary actions (C); nonintentional actions (D), and nonactions (E). (We will assume that all of these actions are free, that is, that no coercive incentives are involved, so that we will not give examples of actions in category B.) It is interesting to note how the lack of category C leads to special problems in describing certain conditions for the authors of DSM-III. Sometimes they solve these problems but sometimes not.

Intentional voluntary actions (*category A*)

Malingering is an example of a voluntary action which sometimes confronts physicians. The person who feigns symptoms does so intentionally and has the volitional ability to will to feign or not to feign. One regards the malingerer as having acted both intentionally and voluntarily and therefore as being fully responsible for his behavior.

DSM-III classifies malingering as a V Code (pp. 331–332), which seems appropriate. Malingerers do not suffer from a mental malady. Malingering is not even a condition of the person; it is the name given to a particular kind of action which is intentionally and voluntarily performed. The action is often quite rational in that the benefits the person may gain are worth the risks of discovery.

DSM-III correctly notes that malingering is a voluntary action: "The essential feature is the voluntary production and presentation of false or grossly exaggerated physical or psychological symptoms" (p. 331). We think it might be preferable to say that malingering involves "the intentional and voluntary production," etc., thereby explicitly noting that the malingerer not only wills to produce the exaggerated symptoms of a malady but also that he has the volitional ability to will to produce them. This allows one to distinguish the malingerer from someone suffering from Münchausen's syndrome (discussed in more detail later in this chapter) very simply; both produce their symptoms intentionally, but the malingerer has the volitional ability to will to produce them and hence does so voluntarily, whereas someone suffering from Münchausen's syndrome does not.

Intentional unvoluntary actions (category C)

There are many behaviors of interest to psychiatrists which fall into this category. These actions all involve the patient's having a volitional disability. *Compulsions* are an obvious example and have been alluded to above. The patient carries out his compulsion intentionally—for example, he wills to wash his hands repetitively—but he does not do so voluntarily since he cannot will to refrain from washing his hands frequently over a period of time. It is, in fact, correct to say that he does not have the volitional ability to will *to* wash his hands frequently over a period of time since having that ability requires refraining from washing one's hands in the presence of appropriate coercive and noncoercive incentives, and this he cannot do.

DSM-III notes that compulsions are "repetitive and seemingly purposeful behaviors" that "are performed with a sense of objective compulsion coupled with a desire to resist the compulsion (at least initially)" (p. 235). We think that the phrase "seemingly purposeful" is not quite right; not only is its meaning unclear, but with some plausible interpretations it is seriously misleading. First, it is not clear if "seemingly purposeful" is to be contrasted to "actually purposeful"; if it is, then "seemingly purposeful" does not seem an accurate description of the behavior of a compulsive hand washer, for his behavior is actually purposeful. Though one might claim that its seeming purpose, keeping his hands clean, is not its real purpose, this idea is not clearly conveyed by the phrase, "seemingly purposeful".

Furthermore, there are other compulsive acts which do not even have a plausible seeming purpose, such as avoiding stepping on cracks in the sidewalk. It is true that avoiding sidewalk cracks, because intentional, may appear more purposeful than such phenomena as tics or twitches or reflex actions. If this distinction is what is meant, then we think the authors should have used the term "intentional." One can see that "seemingly purposeful" is not correct by noting that some behavior, even repetitive behavior, done by someone during, say, an episode of a "Sleepwalking Disorder" is "seemingly purposeful" but is not regarded as a symptom of a compulsion. Compulsive behavior must be intentional behavior. This explains why the "seemingly purposeful" behavior of sleepwalking does not count as compulsive behavior: it is not intentional. It is the intentionality of the behavior, not its seeming purposefulness, that is important. It may be that failing to distinguish between "intentional" and "voluntary" kept the authors of DSM-III from using "intentional" to describe behavior they did not regard as voluntary.

It is important to note that the DSM-III authors, because they lack the concept of a volitional ability, include an initial (conscious?) desire to resist in their characterization of compulsions. This initial desire to resist probably is not really significant. Of course, it is likely that most persons with compulsions have had desires to resist them, but we think that this is not essential. We think that it would make perfect sense to prove to someone that he had a compulsion which he may or may not have recognized because he had never tried to resist it. On our account, one could prove this by providing the person with numerous and varied noncoercive incentives for not acting and noting that none of them were successful in getting him to refrain. Even stronger proof could be provided, if it were morally allowed, by providing, or getting him to believe that we were providing, coercive incentives for refraining and noting that they were also unsuccessful.

We are not claiming that any intentional unvoluntary action is a compulsion; addictions are distinguished from compulsions, and they also involve intentional unvoluntary action. Indeed, we think that our category of intentional unvoluntary action brings out the obvious similarity between compulsions and addictions. The difference between them is that addictions involve substances that we know cause physiological changes in the body, whereas compulsions do not.

Phobias are also frequently associated with intentional unvoluntary behavior. DSM-III defines phobias chiefly in terms of the patient's "irrational fear," but there is some ambiguity and inconsistency about whether the individual must also actually avoid the feared object or situation. The relevant diagnostic criterion for "Agoraphobia" (p. 227) stipulates that the individual *does* avoid being alone or in public places;

for "Social Phobia" (fear of distressing social situations) and "Simple Phobia" (any phobia other than agoraphobia or social phobia; for example, fear of heights) (pp. 227, 228) it is stated only that the individual has a "compelling desire to avoid" the particular object or situation; the Glossary entry on phobia (p. 366) states fairly clearly that only the irrational fear is necessary, not any actual avoidance behavior.

In our account of mental maladies, we pointed out that either inappropriate anxiety or a volitional disability is sufficient to make a condition a malady. This enables us to discover the ambiguity in the present definition of phobia, and perhaps also of compulsion. It is possible for someone with claustrophobia to feel great anxiety when the situation requires him to enter an elevator, but to enter it in spite of the anxiety. In this case, the phobia is a mental malady because of the intense, inappropriate anxiety, but it does not involve a volitional disability, as the person does enter elevators even when the incentives are noncoercive. However, the phobia may also involve a volitional disability: the person may not be able to will to enter an elevator under any circumstances, even in the presence of coercive incentives.

Present theory says that if there is a volitional disability involved in a phobia, it is due to the great anxiety caused by the phobic situation. We do not question this theory, but only wish to point out that it is at least (logically) possible for one to suffer a phobia which involves a volitional disability without suffering any conscious anxiety. We point this out to emphasize that anxiety and volitional disability are separable evils, and that when considering phobias (and perhaps compulsions) it should be made clear whether the phobia includes both of these evils or only one of them, and if so, which one.

Here we shall only refer to the common situation in which the phobic person always avoids the object of his fears, and thus suffers a volitional disability. This avoidance is intentional unvoluntary behavior in that though the phobic person acts intentionally, he does not act voluntarily, that is, though he wills, he does not have the volitional ability to will to avoid the situation. In particular, although the phobic person believes that sometimes there are noncoercive incentives for not avoiding the situation (e.g., he believes that there are sometimes some important benefits in entering an elevator), he never acts on these incentives. Indeed, he often does not act even when there are coercive incentives, such as a fire in the building.

Another interesting example of intentional unvoluntary behavior is found in the "Factitious Disorders." These are disorders in which the patient produces for the physician various symptoms which are either fabricated or self-induced. Because these disorders are not widely known, we will first quote from the DSM-III description of one of them, "Factitious Disorder with Physical Symptoms."

301.51 Chronic Factitious Disorder with Physical Symptoms
The essential feature is the individual's plausible presentation of factitious physical symptoms to such a degree that he or she is able to obtain and sustain multiple hospitalizations. The individual's entire life may consist of either trying to get admitted into or staying in hospitals. Common clinical pictures include severe right lower quadrant pain associated with nausea and vomiting, dizziness and blacking out, massive hemoptysis, generalized rash and abscesses, fevers of undetermined origin, bleeding secondary to ingestion of anticoagulants, and "lupus-like" syndromes. All organ systems are potential targets, and the symptoms presented are limited only by the individual's medical knowledge, sophistication, and imagination. This disorder has also been called Münchausen syndrome.

These individuals usually present their history with great dramatic flair, but are extremely vague and inconsistent when questioned in more detail. There may be uncontrollable pathological lying, in a manner intriguing to the listener, about any aspect of the individual's history or symptomatology (pseudologica fantastica). . . . Individuals with this disorder often eagerly undergo multiple invasive procedures and operations. . . . When confronted with evidence of their factitious symptoms they either deny the allegations or rapidly discharge themselves against medical advice. They will frequently be admitted to another hospital the same day. Their course of hospitalizations often take them to numerous cities, states, countries, and even different continents. Eventually a point is reached at which the individual is "caught" producing factitious symptomatology; he or she is recognized from a previous admission or another hospital, or other hospitals are contacted and confirm multiple prior hospitalizations for factitious symptomatology. (pp. 287–288)

This disorder differs from malingering in two important ways. The first is that while patients who malinger are usually acting rationally (though immorally), patients with Münchausen's syndrome are not. They have no adequate reason for subjecting themselves to the repeated trials of hospitalization, diagnostic tests, and inevitable discovery which they experience. Indeed, they have both coercive and noncoercive incentives for not subjecting themselves to these risks, and yet they do so. The malingerer will not go through with a painful and dangerous operation; he will get well quickly. Indeed, one sometimes discovers malingerers by presenting them with the possibility of such an operation. However operations often do not discourage those who suffer from Münchausen's disorder.

Münchausen's patients differ from malingerers in that their symptom production and presentation, though often intentional, is unvoluntary. DSM-III makes this distinction fairly accurately but clearly has problems in explicitly stating it:

> "Factitious" means not real, genuine, or natural. Factitious Disorders are therefore characterized by physical or psychological symptoms that are produced by the individual and are under voluntary control. The sense of voluntary control is subjective, and can only be inferred by an outside observer.
>
> The judgment that the behavior is under voluntary control is based, in part, on the patient's ability to simulate illness in such a way that he or she is not discovered. This involves decisions as to timing and concealment that require a degree of judgment and intellectual activity suggestive of voluntary control. However, these acts have a compulsive quality, in the sense that the individual is unable to refrain from a particular behavior, even if its dangers are known. They should therefore be considered "voluntary" in the sense that they are deliberate and purposeful, but not in the sense that the acts can be controlled. Thus, in Factitious Disorders, behavior under voluntary control is used to pursue goals that are involuntarily adopted. (p. 285)

Here the manual must invoke two senses of the word "voluntary." Instead, we reserve this term for that sense which carries the psychiatric force and replace DSM-III's misleading use of "voluntary" with the term "intentional." Thus, we would say that Münchausen patients act intentionally but not voluntarily. We think that the phrase "intentional unvoluntary action" has a very useful role to play in psychiatric description.

There are many other instances of intentional unvoluntary behavior of interest to psychiatrists, such as alcoholism, obesity, kleptomania, and many kinds of ego-dystonic sexual behavior (see Chapter 5). Patients suffering from all of these disorders manifest a volitional disability as a central feature of their condition, though this is often overlooked because their actions are intentional.

One last example of some importance is "Noncompliance with Medical Treatment," behavior which DSM-III classifies as a V Code. DSM-III describes the condition in the following way:

> Noncompliance with Medical Treatment
>
> This category can be used when a focus of attention or treatment is noncompliance with medical treatment that is apparently not due to a mental disorder. Examples include failure to follow a prescribed diet because of religious beliefs or to take required medication because of a considered decision that the treatment is worse than the illness. The major differential is with Personality Disorders with prominent paranoid, passive-aggressive, or masochistic features. (p. 333)

DSM-III's examples of noncompliance are of rational, intentional, voluntarily chosen refusals: "failure to follow a prescribed diet because of religious beliefs or to take required medication because of a considered decision that the treatment is worse than the illness." Given these examples, it is not at all clear why a patient's voluntary, rational refusal of a treatment should even be given a V Code, as this suggests that something is wrong and that medical attention might be useful.

However, there is a kind of noncompliance that does represent a mental malady and sometimes a quite serious one. That is when patients do not have the volitional ability to will to comply with treatment prescriptions that they themselves express a strong desire to heed. There are many such examples: diabetics who repeatedly go out of control because of their choice of foods; adolescent epileptics who fail to take their anticonvulsant medications and thereby have frequent seizures; and so on. In these instances, the behavior is irrational (the patients have no adequate reasons for doing that which they know will lead to their suffering significant evils, though they usually have apparent, sometimes unconscious, motives) and the patient acknowledges this fact, yet the behavior persists. These patients have a volitional disability and therefore a mental malady. There should be a classification for such patients in DSM-III, but there is none. This problem was apparently recognized by the writer of the V Code description of "Noncompliance," for the last sentence of the description reads: "The major differential is with Personality Disorders with prominent paranoid, passive-aggressive, or masochistic features." Perhaps some patients who engage in unvoluntary noncompliance suffer from personality disorders, but that point needs to be shown or proved and not just stated. For example, noncompliance is not noted in DSM-III as a defining criterion or an associated feature of either Paranoid or Passive-aggressive Personality Disorders.

Nonintentional actions (category D)

An example of a psychiatrically interesting condition which involves nonintentional actions is accident-proneness. Accident-prone individuals behave in ways that result in nonintentional self-injury. Essentially everyone manifests such behavior on occasion, but when an individual consistently does so with a frequency which seems too high to be due to chance alone, then he is often said to be accident-prone. It is part of the usual theory of accident-proneness that these "accidents" are not just bad luck, though it is acknowledged that they are not intentional. Rather, they are usually attributed to unconscious needs or motives, such as guilt. The condition of accident-proneness is not listed in DSM-III, but it does satisfy the definition of a malady in that accident-prone individuals do have an increased risk of suffering evils. We can see no reason why accident-proneness, at least in its more extreme and obvious form, should not be a DSM-III disorder. Of course, there are many borderline cases in which one is not certain whether to attribute a person's frequent accidents to the person himself or to uncommonly bad luck, so malady status should not be conferred except when there is a clear pattern of accident-causing behavior.

A final example of psychiatrically relevant nonintentional action is a slip of the tongue, which is often referred to as a "Freudian slip." In these cases, a person intends to say one thing but actually says something else, often in basic conflict with the intended utterance. The person nonintentionally uses a word which is inconsistent with the meaning he intends to express but consistent with some second meaning, often opposed to the first, which seems to express attitudes or feelings that we know the speaker has. Freud has provided many examples of such slips, some of them undeniably demonstrating the nonintentional character of these actions.

Nonactions (category E)

Movements made during major motor epileptic seizures represent a clear case of a nonaction. They are independent of the will of the patient, who is unconscious, and they have physical rather than mental causes. "Tourette's Disorder" (DSM-III, 1980, pp. 76–77) is a similar condition occurring in patients who are not unconscious. These individuals have a malady in which they experience "recurrent, involuntary, repetitive, rapid movements [tics], including multiple vocal tics." "The vocal tics include various sounds such as clicks, grunts, yelps, barks, sniffs, and coughs, or words." In about 60 percent of all cases, obscene words are uttered, a condition called "coprolalia" (DSM-III, 1980, p. 76). The vocal tics of Tourette's patients are not merely nonintentional; they do not seem to be actions at all. As we use the term "action," the movement (or lack of movement) must include, at least in part, willing to do something. Most nonintentional actions, such as wounding the picknicker, involve an intentional action, such as shooting the bird. But the symptom's of

Tourette's syndrome, such as the obscene words, involve no willing at all. Thus we do not regard these utterances as actions, in our sense of the term. Similarly, we do not count reflex actions such as knee jerks and eye blinks as genuine actions, but only as bodily movements.

Psychiatrically interesting cases of bodily movements that are not actions may include hysterical seizures. These seizures resemble epileptic seizures, do not seem intentional, and do not seem to involve any willing. They come over the patient in a manner closely resembling that of epileptic seizures. However, we know that there is no brain disorder, so that these bodily movements, though not considered actions, are nonetheless psychologically caused. The existence of such bodily movements independent of the will, but clearly having mental rather than physical causes, reinforces Freud's view about unconscious mental processes. It is also compatible with mental causation for what seem like normal physical disabilities, such as hysterical blindness and hysterical paralysis. In addition, it fits well with mental causation of genuine bodily changes, such as burnlike blisters on the skin from the hypnotic suggestion that a lighted cigarette has been ground into one's palm. The fact that very similar symptoms can have either psychological or physical causes reinforces our view that there is no essential distinction between physical and mental maladies.

Notes

1. For a fuller discussion of voluntary abilities, see Duggan and Gert (1967).
2. For a fuller theoretical discussion of volitional ability, see Gert and Duggan (1979).

References

American Psychiatric Association. *Diagnostic and Statistical Manual of Mental Disorders. DSM-III.* Washington, D.C.: American Psychiatric Association, 1980.

Duggan, Timothy J., and Gert, Bernard. Voluntary abilities. *American Philosophical Quarterly*, 1967, *4*, 127–135.

Gert, Bernard, and Duggan, Timothy J. Free will as the ability to will. *Nous*, 1979, *13*, 197–217.

7

Paternalistic Behavior

Paternalistic behavior is pervasive in medical practice.[1] The issue of when such behavior is justified and when it is not lies at the heart of many dilemmas in medical ethics.

Consider the following behaviors:

Treating a patient without his or her consent
Giving a placebo
Withholding full information about prognosis when initially telling a patient he or she has a malignancy
Committing a psychiatric patient to a state hospital
Performing a surgical procedure without telling a patient that some physicians use an alternative medical one

While these behaviors vary in content, they all might be (and usually are) examples of paternalistic behavior. In this chapter, we give a definition of paternalistic behavior that presents those attributes which all acts of paternalism share. In the next chapter, we will discuss the important issue of when paternalistic behavior is justified and when it is not.

Defining paternalistic behavior: background

Gerald Dworkin (1971), in an important article, defines paternalism too narrowly: "By paternalism I shall understand roughly the interference with a person's liberty of action justified by reasons referring exclusively to the welfare, good, happiness, needs, interests, or values of the person being coerced." All of Dworkin's examples are of laws or regulations which he considers paternalistic. Though he does recognize that there is such a thing as "parental paternalism," he simply assumes that it will always involve the parent's attempt "to restrict the child's freedom in various ways" (p. 119). Paternalism in law doubtless does involve interference with liberty most of the time, but this is due to the nature of law, not to the nature of paternalism.[2] The first of the above quotations also suggests that Dworkin incorrectly regards interfering with a person's liberty of action as entailing coercion of the person.[3] The following example shows that an adequate account of paternalism must allow not only for paternalistic action in which no one is being coerced but also for paternalistic action which does not involve interfering with anyone's liberty of action.

Case 7-1.

Mr. N, a member of a religious sect that does not believe in blood transfusions, is involved in a serious automobile accident and loses a large amount of blood. On arriving at the hospital, he is still conscious and informs the doctor of his views on blood transfusion. Immediately thereafter he faints from loss of blood. The doctor believes that if Mr. N is not given a transfusion he will die. Thereupon, while Mr. N is still unconscious, the doctor arranges for and carries out the blood transfusion.

The doctor's action here seems to be clearly paternalistic and does in fact satisfy the definition we give below. The example shows not only that paternalistic action need not be coercive and need not involve an attempt to interfere with a person's liberty of action but also that it need not even involve an attempt to control the person's behavior. Coercive action, which involves the use of threats, is a subclass of attempts to interfere with liberty of action. Attempts at such interference are, in turn, a subclass of attempts to control behavior. Thus, by showing that we can have paternalistic action which does not involve an attempt to control behavior, we can show that paternalistic action need not be coercive or involve an attempt to interfere with liberty of action. In the blood transfusion case there was no attempt to control behavior; indeed, there was no behavior to control. Thus it seems clear that there was no attempt to interfere with liberty of action and no coercive action.

The same points can be made by considering an example of paternalistic deception which is intended to affect feelings rather than behavior. Suppose a doctor lies to a mother on her deathbed when she asks about her son. The doctor tells her that her son is doing well, although he knows that the son has just been killed trying to escape from prison after having been indicted for multiple rape and murder. The doctor behaved paternalistically but did not attempt to control behavior, to apply coercion, or to interfere with liberty of action. Even in political rather than personal situations, paternalism may involve deception in order to affect the body rather than behavior—for example, officials who surreptitiously introduce fluorides into a city's water supply in order to reduce tooth decay in the inhabitants.

Of course, many paternalistic acts do involve attempts to control behavior, but even these are not all best described as attempts to deprive a person of freedom.[4] The following is a clear case of paternalism involving the deprivation of freedom.

Case 7-2.

Mr. K is pacing back and forth on the roof of his five-story tenement and appears to be on the verge of jumping off. When questioned by the police, he sounds confused. When interviewed by Dr. T in the emergency room, Mr. K admits to being afraid that he might jump off the roof and says that he fears he is losing his mind. However, he adamantly refuses hospitalization. Dr. T decides that for his own protection, Mr. K must be committed to the hospital for a period of forty-eight hours.

The following case, though it might be described as deprivation of freedom, seems more accurately characterized as depriving a person of opportunity.

Case 7-3.

Professor M tells his wife that he has had a brief affair with her best friend. On hearing this, his wife becomes very depressed and says that she wants to kill herself. In fact, she once took an overdose of sleeping pills when she was depressed. Before leaving for a class that will be over in two hours he, without telling her, removes all the sleeping pills from the house.[5]

There are other paternalistic acts that involve physically disabling a person; for example, a mother, convinced that her son will be killed if he is drafted, breaks his arm in order to prevent that from happening. This somewhat unusual case may be described as paternalistic but is not best characterized as depriving the son of freedom, especially if he has no desire to be drafted. Knocking someone out by a blow, as one might do to a friend who seemed about to attack an armed robber, can be a paternalistic act, but it is more plausibly described as temporary disabling than as deprivation of freedom. Disabling the will of a person, that is, doing something that takes away his volitional ability to will to do certain kinds of acts, can also be paternalistic. Thus, certain kinds of aversive conditioning may result in a lack of a specific volitional ability to will, and may be done primarily to prevent to patient from carrying out actions which are harmful to himself.

All of the paternalistic acts described above involve doing something that needs moral justification. This is because an essential feature of paternalistic behavior toward a person is the violation of one or more moral rules. We discuss moral rule violations in more detail below, but all of the paternalistic actions described above involve violating such moral rules as those prohibiting deception, deprivation of freedom or opportunity, or disabling. One can even imagine a paternalistic act that involves breaking the moral rule against killing, for example, killing a person who has developed the symptoms of rabies and thus faces a certain and excruciatingly painful death. Obviously, one can also act paternalistically by violating the moral rule prohibiting the causing of pain, either physical pain or mental suffering, if, for example, it is done in order to prevent what one believes to be even greater pain or suffering, as in the following case.

Case 7-4.

Mrs. B will undergo surgery in two or three days for a malignant tumor of her right breast. She has obviously understood her situation intellectually, but her mood has been rather blasé and she appears to be inappropriately minimizing the emotional gravity of her situation. Dr. C's experience is that women in Mrs. B's situation who before mastectomy do not experience some grief and at least moderate concern about the physical and cosmetic implications of their operation often have a very severe and depressive postoperative course. Though Mrs. B has insisted that she does not wish to talk about the effects of the surgery, Dr. C talks with her about such effects before surgery in order to facilitate her emotional preparation for her impending loss.

This last example also goes against a common view of medical paternalism. If one is presented with the question "which doctor is acting paternalistically, one who confronts a patient with a painful truth, or one who withholds the truth in order to avoid the pain it will cause the patient?" most will choose the latter. But as the example above makes clear, this need not be the case. Which doctor is acting paternalistically, if either one is, depends upon whether he will proceed with what he thinks is best for the patient regardless of the patient's expressed wishes. If the patient wants to be told the truth, then to withhold it simply to prevent his suffering the effects of being told is paternalistic. But if the patient says that he does not want to be told the truth—say, about his having terminal cancer—then it is paternalistic of the doctor to cause suffering by forcing the truth on the patient on the grounds that it is better for him to face the painful truth now.

The definition of paternalistic behavior

With this background, we offer the following definition of paternalistic behavior:

A is acting paternalistically toward S if and only if (A's behavior correctly indicates that *A believes that*):[6]

1. his action benefits S
2. his action involves violating a moral rule with regard to S
3. his action does not have S's past, present, or immediately forthcoming consent
4. S is competent to give consent (simple or valid).

From this definition, it is easy to derive accounts of paternalistic attitudes, persons, laws, and so on, but we shall not consider these matters here. What we wish to do now is to discuss the various features of our definition.[7] There is no dispute about (1). If A is acting paternalistically toward S, then A must intend to benefit S, not himself or some third party. This is not to deny that actions can be partially paternalistic; they can be intended for the benefit of S and others, including A. But what makes A's actions toward S paternalistic is never the intended benefit of anyone other than S himself.

There are two ways in which a person can act to benefit a second person. The first is to procure some good for the second person, for example, to increase another's pleasure, freedom, or opportunity. The second is to prevent or ameliorate some evil the second person might suffer, such as pain or disability. Paternalistic acts can involve benefiting of either kind although, as we will discuss in the next chapter, it is only the paternalistic prevention of evils for another which is sometimes justified.

Feature 2 involves an expansion of what is normally said about paternalism, namely, that it involves a deprivation of freedom of the person toward whom one is acting paternalistically. We have seen that paternalism need not always involve violating the moral rule prohibiting the deprivation of freedom or opportunity, and that it can involve violating the moral rules against killing, causing pain, disabling, depriving of pleasure, deception, or breaking a promise. Paternalism involving breaking a promise is even discussed by Plato, who advocates not returning a weapon to someone who has gone mad, even though one has promised to do so.

Violating a moral rule involves doing something that would be morally wrong unless one has an adequate justification for doing it. Thus, killing, causing pain (mental or physical), disabling, and depriving of freedom, opportunity, or pleasure are all violations of moral rules. The same is true of deceiving, breaking a promise, and cheating. All of these actions are such that one is *never* morally allowed to perform them simply because one wants to do so. Every instance of these actions is morally prohibited unless a morally adequate reason is provided. Though there may be disagreement on what counts as a morally adequate reason, there is no disagreement that performing the actions under discussion is immoral unless one has such a reason. We use the phrase "violation of a moral rule" to refer to such actions.

Paternalistic behavior always involves the belief (or knowledge) that one is performing one of these actions; it does not require one to think of these actions as violations of moral rules. It is customary to talk about killing, deceiving, and breaking a promise as violations of moral rules, but there is no such linguistic tradition with regard to causing someone mental or physical pain or disabling him, although to do so without adequate justification is clearly immoral. We see no reason for distinguishing killing and deceiving from causing pain and disabling, and we think it most fruitful to regard all of these acts, as well as depriving of freedom or opportunity, as violations of moral rules.[8]

We do distinguish between violating a moral rule and failing to follow a moral ideal. Following a moral ideal is helping someone in need, for example, preventing or relieving pain. We talk about the failure to follow a moral ideal when there is an occasion to follow it and one fails to do so, for example, refusing to give a beggar any money. We distinguish violations of moral rules from failure to follow moral ideals because we hold that the former always need moral justification and the latter do not. Since we hold that paternalistic behavior always requires justification we restrict the application of that phrase to violations of moral rules and do not count failure to follow moral ideals as paternalistic. Thus we do not count it as paternalistic behavior if one refuses to give money to a beggar because one believes he will only buy whiskey with it and that will be harmful to him. Such behavior may reveal a paternalistic attitude, that is, a willingness to act paternalistically toward the beggar if the situation arose, but it is not itself a paternalistic act. It is only when one's action requires moral justification that we think it appropriate to call it paternalistic. We realize there will be some disagreement concerning what acts need moral justification, but we believe that there is agreement on the overwhelming majority of cases.

Feature 3 makes it clear that A believes that his violating a moral rule for S's benefit does not have S's past, present, or immediately forthcoming consent. If A has S's consent, or if A expects S's immediately forthcoming consent for his action, then an action which might otherwise by paternalistic is not so. For example, suppose I pull someone from the path of an oncoming car which I believe he does not see. If I act because I think that he will approve of my action immediately, my action is not paternalistic even though it may satisfy all the other conditions of paternalistic behavior. But if I think that he is trying to commit suicide because of a temporary depression and that he will thank me later when he recovers, then my act *is* paternalistic. Further, suppose I prevent someone from signing a contract because I believe he does not know that he is being swindled. Then, even though I think he will thank me when he learns the facts, if I cannot present those facts immediately, I am acting paternalistically.

However, past, present, or immediately forthcoming consent removes A's act from the class of paternalistic acts. For example, suppose a very distraught patient goes to a psychiatrist and says that he knows that he badly needs treatment and would like the psychiatrist's advice about whether to be hospitalized. The psychiatrist is not acting paternalistically if he urges the patient to enter the hospital.[9] But supoose that, later, the patient wishes to be discharged from the hospital, and the psychiatrist refuses permission because he thinks the patient may do himself serious harm if he is allowed to leave. The psychiatrist's refusal is a paternalistic act.

The expectation of receiving future consent is sometimes present for A when acting paternalistically, but it is important to note that such an expectation, even when a virtual certainty, does not make an action nonpaternalistic. For example, a psychiatrist may hospitalize a depressed suicidal patient against the patient's wishes, and the circumstances of the case may be such that the psychiatrist knows almost for certain that the patient will be effusively thankful within two or three days. The hospitalization nonetheless is a paternalistic action; whether or not it is justified will be discussed in the next chapter.

Feature 4 is presupposed in many accounts of paternalism but rarely is made explicit. We can be paternalistic only toward those whom we regard as competent to give either simple or valid consent. Thus, we cannot act paternalistically toward infants and comatose persons because they are not competent to give even simple consent.

It is important to note the way in which this fourth feature of the definition is related to our earlier discussion, in Chapter 3, of patients' competence. Any patient who is believed by his physician to be competent to give either simple or valid consent to treatment would satisfy this condition. Frequently, it might be true that patients who were competent to give only simple consent (not valid consent) would be believed by their physician to be mistaken about what was for their benefit. This would be the case, for example, if a woman irrationally believed she was not ill and therefore refused to consent to treatment. On our account, she would be incompetent to refuse valid consent (see Chapter 3), but she would be competent to refuse simple consent, and any action directed toward her which also satisfied the first three features of the definition would be paternalistic and would require moral justification. However, patients who are incompetent to give even simple consent, like infants, do not satisfy it, and thus therapeutic actions directed toward them are not paternalistic.[10]

Other definitions of paternalism

Several other definitions of paternalistic behavior have been suggested and we will examine several of them here. We have already criticized Dworkin's (1971) definition as too narrow, at the other extreme, the broadest of the other definitions has apparently been N. Fotion's (1979) which, like the dictionary definition we discussed in Chapter 1, applies "paternalism" to *any* behavior typically engaged in by a father toward a child. Fotion explicitly includes the "providing of services to one's children [which] involves no violation of a moral rule" (p. 195). But the word "paternalism" is simply not used in the way he suggests and there seems to be no advantage in redefining the term in a new way that would now include many kinds of behavior which require no moral justification, such as giving food to one's hungry child. It is the need for justification that makes paternalism an interesting and important moral concept and it is this feature that we think we capture better than alternative definitions.

We think that Fotion has failed to recognize that "paternalism" suggests a distortion of paternal behavior, just as "scientism" suggests a distortion of scientific behavior. But unlike "scientism," "paternalism" allows for this suggestion to be eliminated; there is justified paternalism. But all standard uses of "paternalism" are such that it applies only to behavior which needs justification.

Both Buchanan (1978) and Childress (1979) offer definitions of paternalistic behavior that *are* concerned with actions which require moral justification. Their definitions differ from ours in the way they describe the class of behaviors that need justification. Buchanan begins, as we do, by criticizing Dworkin's definition for being too narrow and then gives the following expanded definition:

> . . . paternalism is interference with a person's freedom of action or freedom of information, or the deliberate dissemination of misinformation, where the alleged justification of interfering or misinforming is that it is for the good of the person who is interfered with or misinformed. (p. 372)

This definition does enlarge Dworkin's definition by adding deception as a type of moral rule violation that may be committed paternalistically, but it is still too narrow with regard to the range of possible moral rule violations. For example, consider the case of Mrs. B (7-4), whose doctor insisted, against her wishes, that she emotionally confront the possible impending surgical loss of her breast. His action satisfies all four features of our definition, but he violated the moral rule prohibiting the causing of physical or psychological pain. He did not interfere with Mrs. B's freedom of action, nor did he deceive her; thus, his behavior is not paternalistic by Buchanan's definition. One could also construct a case in which a promise was broken paternalistically. Thus, while Buchanan's definition is an improvement on Dworkin's, it is still too narrow. It is also true that Buchanan's definition (like Dworkin's) does not explicitly mention the important factor of S's lack of consent (the third feature in our definition).

James Childress (1979) offers a definition of paternalism that has two features:

> The first feature . . . is the motivation and intention to prevent harm to or to benefit another person. (p. 18)
>
> . . . a second feature: it involves a refusal to accept or to acquiesce in an individual's choices, wishes, and actions. . . . It usurps a right of decision-making on the grounds that someone else can make better decisions. (p. 19)

Childress's second feature presents problems. One is that the two sentences that define it have quite different meanings, whereas Childress apparently intends the second sentence merely as a restatement or amplification of the first. For example suppose a hypochondriacal patient requests that a surgeon perform an operation on him which the surgeon feels is unnecessary and contraindicated. If the surgeon declines, then this "involves a refusal to accept or to acquiesce" in the patient's choices or wishes but it does not "usurp a right of decision-making" from the patient, since the patient has no right to require the surgeon to operate. Which sentence should we use? The first sentence, by itself, does not work as a statement of Childress's second feature. If the surgeon refuses because he believes it in the patient's benefit not to have unnecessary surgery, then he has satisfied both features of Childress's definition, but surely his refusal to operate is not a paternalistic act. For Childress (p. 19), as well as for us, paternalistic behavior always requires moral justification, but one need not morally justify refusing to do unnecessary surgery.[11]

Using Childress's second sentence (". . . usurps a right of decision-making . . .") as a statement of the second feature does not work either. It is not clear, first of all, what the sentence means: for example, does it refer to not allowing someone to make a decision in the first place, or to allowing him to make it but not carry it out? Whatever is meant, the sentence strongly suggests that all paternalistic acts are done to affect someone's decision-making in some way, which is not true, as we have shown by providing the example of the doctor lying to the mother on her deathbed about her son. One might counter by trying to define "usurping the right of decision-making" in such a way that it covers all acts of deception. We believe, however, that taking "usurping the right of decision-making" so broadly that it covers deception and all other actions requiring moral justification makes it mean no more than simply violating someone's rights. As we stated earlier, we think the concept of moral rule violation is clearer than that of rights violation, but even if one prefers the rights terminology it is clear that more than the right of decision-making is involved.

These various disagreements about the appropriate definitions of paternalism are not just hair splitting. For example, if one holds, as most authors on the topic do, that paternalistic behavior needs to be justified, then it is important that one's definition not mistakenly label some behaviors as paternalistic (e.g., feeding one's children or refusing to perform unnecessary surgery), and thus in need of justification, when there is no such need.

Applications of our definition

Now that we have given our definition of paternalistic behavior and explained it in some detail, let us consider some cases of psychiatric intervention and see whether the account is of any help in determining whether they are paternalistic or not.

Case 7-5.

Miss X is a twenty-two-year-old single woman brought to the Emergency Department by her roommate, who accidentally came home and discovered her soon after Miss X ingested what appears (and later proves) to be a lethal quantity of barbiturate capsules. Miss X's reason for doing this is that her fiancé has been killed the day before in an automobile accident and she feels she does not want to live without him.

Dr. W tells Miss X that it will be necessary to insert a naso-gastric tube in order to evacuate the contents of her stomach. She refuses permission, saying that she wishes to die. Dr. W nonetheless inserts the tube, despite Miss X's angry objections, and recovers a large quantity of the capsules.

Dr. W's action of inserting the naso-gastric tube intuitively seems paternalistic and in fact does satisfy the features of our definition:

1. Dr. W believes that his action is for Miss X's benefit, in that it will very likely prevent her death or serious injury (e.g., brain damage from respiratory depression resulting in lack of oxygen in the brain).

2. Dr. W believes (knows) that his actions deprive Miss X of freedom and cause her pain, both of which count as violations of moral rules.
3. Dr. W believes that he does not have Miss X's past, present, or immediately forthcoming consent; in fact he knows that she has refused consent.
4. Dr. W believes that Miss X is competent to give or refuse valid consent. She understands and appreciates her current situation, knows the consequence of her refusal of consent, and knows that she is being asked to give consent.

Let us consider another example, this time involving deception:

Case 7-6.

Mrs. P, on her first visit as an out-patient, is insistent during the last few minutes of her session that Dr. Z give her some medicine for her nerves and for the vague, poorly localized pains which she describes. He feels there is no medical reason for her to have medication but judges that if he refuses her request outright, a useful and productive initial interview will end on a very sour note. However, he believes strongly in not administering active drugs when there is no medical reason for doing so; therefore, he writes a prescription for a week's supply of a placebo and makes a note on her chart to discuss the issue of medication with her in detail at their next appointment.

Dr. Z's act is clearly paternalistic. He has given her a placebo for what he believes is her benefit; he knows he is deceiving her and that he does not have her consent (in this case, even her knowledge); and further, he views her as someone who is competent to give valid consent, that is, as one who understands and can appreciate all of the information given to her about her condition.

Let us consider an example that is somewhat less straightforward:

Case 7-7.

Dr. Q is leading a new therapy group during its second session. The group consists of patients who have all claimed to have difficulty in relating to other people.

One patient, Mr. G, is a single professional man in his early thirties who has complained of an inability to maintain lasting friendships with either men or women. It has become apparent to Dr. Q through watching the group interaction that Mr. G, while not totally unlikeable, is self-centered, critical of others, and smugly certain about his own opinions. It has also become apparent that Mr. G has little insight into these characteristics and the way they irritate other members of the group. Dr. Q believes it would be useful for Mr. G to acquire insight into the effect his personal style has on others. Of course, whether Mr. G will then try to change his style will be his own decision.

Accordingly, midway through the session, Dr. Q begins to encourage other group members to confront Mr. G with their feelings about him, despite Mr. G's obvious anger and great discomfort when they begin to do so.

This satisfies three of the four elements of the definition. Dr. Q believes that confrontation is for Mr. G's benefit, knows that he is causing Mr. G mental suffering, which is violating a moral rule, and thinks that Mr. G is someone who is competent to give valid consent to being confronted. What is not clear from the example as given is whether Dr. Q believes he has Mr. G's consent to confront him. This is not clear because nothing is specified about the nature of the prior agreement between Dr. Q and Mr. G. In particular, nothing is said about whether Mr. G has or has not consented to confrontation-type activities (i.e., experiencing emotional pain in the hope of achieving greater self-understanding). Thus, on the information given, we do not know whether Dr. Q's actions are paternalistic. Dr. Q might claim that Mr. G's presence in group therapy demonstrates implicit consent to being confronted in an emotionally painful way, but that claim would be weak unless Dr. Q had evidence that essentially *all* patients had such expectations, which seems unlikely.

We stress this example because it seems similar to the dilemma posed by many psychiatric interventions: whether they are paternalistic (and thus require justification) turns heavily on the nature and quality of the consent given by the patient and the degree to which the psychiatrist acts independently of that consent for what he feels is the patient's good.

In this case, if Dr. Q has obtained Mr. G's consent to be exposed to emotionally painful experiences, then Dr. Q's actions are not paternalistic. If this consent has not been obtained, then he is acting paternalistically.

Note that in the above examples the evils perpetrated by A, and (possibly) prevented for S, vary considerably in their intensity. Health professionals in particular should be sensitive to the pervasiveness of paternalistic actions that are concerned with relatively minor evils. Not allowing a patient to nap during the daytime so that he sleeps better at night can be a thoroughly paternalistic action and in as much need of justification as depriving someone of extensive liberty. We will return to this point in the next chapter.

We think that this account clarifies many of the features of paternalism. It explains why doctors, nurses, and other health professionals who act to benefit their patients usually resist the charge of paternalism: people generally do not want to be in a position of having to justify their actions. Thus, even though our account does not allow paternalism to degenerate into a term of abuse—for there can be justified paternalism—it is clear that paternalism involves violating a moral rule without consent and thus requires justification. We can also understand why the concept of paternalism is so closely tied to the notion of valid consent. The doctor-patient relationship demonstrates this point well, for it is often not clear whether and to what extent the patient has given valid consent to the doctor. This vagueness is evident in the example of the patient in group therapy.

In addition to clarifying the concept of paternalism, we think that the description we have provided allows for some interesting empirical research. For example, what factors in medical training lead a doctor to act paternalistically? To what extent would awareness that one was acting paternalistically (as defined in this chapter) decrease (or increase) one's tendency to do so? Is there a significant difference between doctors who often act paternalistically, and those who do so infrequently, with regard to their belief about whether people generally make incorrect decisions affecting their own welfare?

Notes

1. This chapter is adapted from Gert and Culver (1976).
2. Dworkin's view that paternalism always involves the restriction of liberty is the standard one. See, for example, Bayles (1974) and Regan (1974).
3. For a fuller account of coercion, see Pennock and Chapman (1972), especially Chapters 3 and 4.
4. We think it is useful to distinguish between the different ways that one can attempt to control another's behavior. We shall use "depriving of freedom" as

referring to the same kinds of actions as Dworkin's phrases, "interfering with liberty" and "restricting freedom," and distinguish these from other ways of controlling behavior, for example, disabling a person.

5. In this example, whether we regard Professor M's act as paternalistic or not depends upon whether we regard him as depriving his wife of the opportunity to take the sleeping pills. Similar problems arise with regard to companies that are called paternalistic. Suppose a company puts a significantly larger percentage of money per employee into pensions and benefits than most companies do, and a correspondingly smaller one into salaries and wages. Whether we regard the company as paternalistic will depend on whether we see it as depriving its employees of the opportunity to spend their money as they choose. If we think the company is paying a fair wage as well as putting more than is required into the pension fund, then it is not correct to view the company as paternalistic. It is often difficult to determine the difference between depriving someone of something and simply not providing him with that thing. We make no attempt in this chapter to resolve this problem. We do claim that an action should be characterized as paternalistic only when it deprives someone of an opportunity (or in some other way violates a moral rule). We wish to thank Gerald Dworkin for calling our attention to this problem.
6. The parentheses in this sentence are meant to indicate that while one may describe A's behavior as paternalistic without explicitly knowing A's beliefs, A's behavior counts as paternalistic only if A does have the relevant beliefs. Further, throughout our analysis, we assume that A's beliefs are at least plausible, though they need not be true. If A's beliefs are wildly false—for example, if he thinks flowers are competent to give valid consent—then we may hesitate to maintain that he is acting paternalistically toward the flowers when he waters them though he believes that they would prefer to remain dry. We are indebted to Timothy Duggan for calling our attention to this latter point.
7. This is a different version of the definition from the one we presented in earlier publications (e.g., Gert and Culver, 1976), though there are no substantive changes. In addition to the parentheses change in the first sentence, we have eliminated our former feature 2 (A believes that he is qualified to act on S's behalf) because we think that A must have that belief in all cases in which he believes the remaining features of the definition, and thus this feature was superfluous. We have also simplified feature 3, removing any mention of justification. Finally, we have simplified our former feature 5 (A believes that S believes (perhaps falsely) that he (S) generally knows what is for his own good) into the current feature 4. The current phrasing is consistent with the account of competence developed in Chapter 3 and more precisely expresses the idea intended by our earlier version.
8. For a fuller account of moral rules, see Gert (1975). If one prefers the language of rights to that of moral rules, one might plausibly hold that all paternalistic behavior involves the violation of a person's rights. The close

connection between rights and liberties may then partly explain the widely held but mistaken view that paternalism always involves the restriction of liberty of action. For example, in the blood transfusion case, it is plausible to say that we are violating the patient's rights, but it is not plausible to say that we are restricting his liberty of action. Similarly, paternalistic behavior involving deception may sometimes be taken as violating the person's right to know when it cannot be taken as restricting his liberty of action. There may be no substantive difference between violating a moral rule with regard to someone independently of his past, present, or immediately forthcoming consent and violating his rights. However, since we find the terminology of moral rules to be clearer than that of rights, we have presented our analysis of paternalism solely in terms of violating a moral rule.

9. We might call this paternal behavior and distinguish it from paternalistic behavior in that it does not involve violating a moral rule with regard to S independently of S's past, present, or immediately forthcoming consent. We do not intend this statement to be taken as a complete account of paternal behavior.

   It should be noted that persons can disagree as to whether behavior is paternal or paternalistic (see note 5). Also, we recognize that the term "paternalistic" is often used to describe behavior that we think is more appropriately described as paternal, but we do not think this is significant to our analysis. (In a similar way, the analysis of jealousy is not significantly affected by the fact that "jealous" is often used to describe an attitude which is more appropriately described as envious.)

10. Feature 4 may be suspended in one very unusual kind of case, that of killing a severely defective neonate in order to prevent it from the great suffering it will experience due to its severe defects. Note that it does not seem paternalistic to take action to save the life of the neonate even if this involves causing considerable pain. We think that killing the neonate to prevent his pain seems paternalistic, whereas causing him pain to prevent his death does not, because the former prevents the neonate from ever coming to be a person, whereas the latter does not. Thus the only time one can be paternalistic toward someone who is incompetent to give simple consent is when one's act prevents that being from developing into a person who would be competent to give simple consent. This is a very special case, and we shall not consider it any further here.

11. This is because not performing unnecessary surgery is not the same as depriving a patient of surgery, and thus no moral rule is violated. The same point can be made about refusing to give money to a beggar, when one does so for the beggar's good (cf. note 5 above). Does the same argument apply to refusing to accede to a patient's request for antibiotics to treat a viral illness, if one refuses for the patient's benefit (for example, to avoid possible allergic reactions)? We believe it does, though some argue that patients have a right to free access to all drugs and that laws requiring prescriptions are therefore themselves paternalistic. But, given these laws, the physician is not acting

paternalistically in refusing to prescribe what he believes might be harmful drugs. Even if there were some dispute about this case, no one would argue that patients have a right to request and obtain any surgery they desire.

References

Bayles, Michael D. Criminal paternalism. In J. Roland Pennock and John W. Chapman (eds.), *The Limits of Law—Nomos XV*. New York: Lieber-Atherton, 1974, pp. 174–188.

Buchanan, Allen. Medical paternalism. *Philosophy and Public Affairs,* 1978, *7*, 370–390.

Childress, James F. Paternalism and health care. In Wade L. Robison and Michael S. Pritchard (eds.), *Medical Responsibility*. Clifton, N.J.: Humana, 1979, pp. 15–27.

Dworkin, Gerald. Paternalism. In Richard Wasserstrom (ed.), *Morality and the Law*. Belmont, Cal.: Wadsworth, 1971, pp. 107–126.

Fotion, N. Paternalism. *Ethics*, 1979, *89*, 191–198.

Gert, Bernard. *The Moral Rules*. New York: Harper Torchbooks, 2nd paperback ed., 1975.

Gert, Bernard, and Culver, Charles M. Paternalistic behavior. *Philosophy and Public Affairs*, 1976, *6*, 45–57.

Pennock, J. Roland, and Chapman, John W. (eds.). *Coercion—Nomos XIV*. New York: Lieber-Atherton, 1972.

Regan, Donald H. Justifications for paternalism. In J. Roland Pennock and John W. Chapman (eds.), *The Limits of the Law—Nomos XV*. New York: Lieber-Atherton, 1974, pp. 189–210.

8

The Justification of Paternalistic Behavior

Introduction

We have stated that all paternalistic behavior requires justification because it involves violating a moral rule with respect to another person without having that person's consent. Logically, there are only three positions that can be maintained with respect to the justifiability of paternalistic behavior: it is always justified, it is never justified, and it is sometimes justified.[1]

Always justified

No current writer on paternalism seems to believe that paternalistic actions are always morally justified. According to our definition of paternalistic behavior, this may seem a plausible position. It is probably true that nearly all persons who act paternalistically *believe* that they are justified in doing so. For a belief that even though one is breaking a moral rule with regard to S without his consent, one is still benefiting him, is required before one's behavior can be considered paternalistic.

It is interesting to note that the ethical theory of act-utilitarianism seems to imply that all paternalistic behavior based on true factual beliefs is justified. That theory holds that in any situation, a person can determine what action is right by choosing that action which will, on balance, cause the most good (and/or prevent the most evil). This is the type of ethical theory that underlies what is sometimes called "situation ethics."

Since this theory denies the significance of moral rules, all that is required to justify one's action is that the amount of good produced by one's action is greater, or the amount of evil less, than would have resulted if one had not acted. It may be that the implicit holding of this mistaken ethical theory is responsible for some paternalistic behavior.

Never justified

The view that paternalistic behavior is never justified has been advanced by Tom L. Beauchamp (1977). Such a position at first seems as implausible as believing that paternalistic behavior is always justified, since examples readily come to mind in which paternalistic behavior seems obviously and strongly justified: for example, preventing a friend, who has become temporarily despondent following ingestion of a psychedelic drug, from killing himself. However, Beauchamp maintains that in such instances we are not really acting paternalistically because our drugged friend's behavior is in some sense not voluntary, and therefore our stopping him is not an act of paternalism. Thus, even someone as antipaternalistic as Beuachamp would presumably agree that some behaviors that we and others would call paternalistic are justified; he would just relabel them as not paternalistic (Beauchamp, 1977, p. 76).

This may sound like a merely verbal dispute. It is true that it is a dispute about the proper use of a word, but it can have significant practical consequences. If one defines paternalism broadly, as we do, so that *any* violation of a moral rule done for a person's benefit but without his consent counts as paternalistic, then one leads anyone who acts in this way to most seriously consider whether his act is justified.

Beauchamp's proposal shifts the emphasis from the genuine moral problem of justifying intervening for a patient's benefit without his consent, to the problem of determining when a person's behavior is nonvoluntary, so that interference is not paternalistic. On our account, even if the patient's behavior is nonvoluntary, it still may not be justified to intervene, for the harm we prevent by interfering may not be great enough to justify our violation of a moral rule with regard to the patient. In Beauchamp's proposal there seems to be an absolute dichotomy between the way we can treat two classes of patients. With regard to those whose actions are voluntary, we can never interfere for their own benefit without their consent, no matter how great the harm we prevent and how minor the violation of the moral rule. With regard to those patients whose acts are not voluntary, we justify our interference by determining that we will prevent harm (Beauchamp, 1977, p. 71). We do not think that there is such a sharp line separating voluntary from nonvoluntary behavior, and, even if there were, it has not yet been reliably enough determined to allow it to play such an important role in determining a doctor's behavior.

An examination of some of the details of Beauchamp's position will clarify these problems. Beauchamp supplies no criteria for determining whether an act is voluntary or not (and therefore, whether an attempt to interfere with the act is or is not paternalistic). He mentions "subliminal advertising . . . drug therapy . . . alcoholic stimulation, mob-inspired enthusiasm, retardation, and psychotic compulsion" (p. 75) as examples of conditions which may cause nonvoluntary behavior. Does a jilted, suicidal teen-ager have a "psychotic compulsion" to kill himself? Surely not every action carried out under "alcoholic stimulation" is nonvoluntary; how are we to decide which are and therefore may be interfered with? Are no actions resulting from an irrational desire to harm oneself voluntary?

A second problem is that Beauchamp gives no guidelines for determining just when we *may* justifiably coerce someone whose nonvoluntary action will result in self-harm. (Beauchamp, like Dworkin, seems to regard all paternalistic behavior as coercive.) Surely one isn't *always* justified in interfering with another person's nonvoluntary actions whenever they result in self-harm. So, even if one were to follow Beauchamp's definitional suggestion, the problem would still remain of when and when not to interfere with a person's nonvoluntary action to harm himself. Beauchamp does not address this problem, surely one of the most vexing that physicians encounter.

Sometimes justified

Many writers maintain, as we do, that some paternalistic behavior is justified and some is not. We will first present our own procedure for justification and then discuss one other procedure which is sometimes used by physicians.

The justification of paternalism

Let us begin by considering two cases of paternalistic behavior, one justified and one not, and introduce some of the factors that are important in the justification process.[2]

Case 8-1.

Mr. K was brought to the emergency room by his wife and a police officer. Mrs K had confessed to her husband earlier that evening that she was having an affair with one of his colleagues. He had become acutely agitated and depressed and, after several hours of mounting tension, told her he was going to kill himself so "you'll have the freedom to have all the lovers you want." She became frightened and called the police because there were loaded guns in the house and she knew her husband was an impulsive man.

In the emergency room Mr. K would do little more than glower at Dr. T, his wife, and the officer. He seemed extremely tense and agitated. Dr. T decided that for Mr. K's own protection he should be hospitalized but Mr. K refused. Dr. T therefore committed Mr. K to the hospital for a seventy-two hour emergency detention.

How could Dr. T attempt to justify his paternalistic commitment of Mr. K? He could claim that by depriving Mr. K of his freedom, there was a great likelihood that he was preventing the occurrence of a much greater evil: Mr. K's death, or serious injury. Dr. T could further claim that in his professional experience the overwhelming majority of persons in Mr. K's condition who were hospitalized did subsequently recover from their state of agitated depression and acknowledge the irrational character of their former suicidal desires.

Note that Dr. T need not claim that self-inflicted death is an evil of such magnitude that paternalistic intervention to prevent it is always justified. Rather, he could claim that it is justified in Mr. K's case on several counts. First, Mr. K's desire to kill himself seems irrational; while one may explain his desire, using psychodynamic concepts, he appears to have no adequate reason for killing himself. An adequate reason would be a belief on his part that his death would result in the avoiding of great evil(s) or the attaining of great goods for himself or others. (His statement to his wife ["You'll have . . . all the lovers you want"] sarcastically expresses an altruistic reason which, even if taken literally, would not be an adequate reason for his suicide.) Second, there is evidence that he suffers from a condition that is well known to be transient. Third, the violation of a moral rule (deprivation of freedom) which Dr. T has carried out results in Mr. K's suffering a much lesser evil than the evil (death) which Mr. K may perpetrate on himself.

But it is not sufficient justification for Dr. T merely to show that the evils prevented for Mr. K by his paternalistic action outweigh the evils caused to Mr. K; he must also be willing to publicly advocate the deprivation of Mr. K's freedom in these circumstances. That is, he must be willing to advocate to all rational persons that in these circumstances everyone may be deprived of his freedom for a limited period of time. (See Gert, 1975, pp. 88–91, for a fuller discussion of public advocacy.) We believe that, accepting the case as described, not only could Dr. T advocate to all rational persons that Mr. K may be deprived of his freedom but also that, if they accepted the facts of the case and the conditions of public advocacy, all rational persons would agree with Dr. T's judgment completely. Thus, Dr. T's paternalistic behavior is strongly justified. (In Chapter 9 we consider the justification of commitment in much greater detail.)

By contrast, consider the following case:

Case 8-2.

Mrs. R, a twenty-nine-year-old mother, is hospitalized with symptoms of abdominal pain, weight loss, weakness, and swelling of the ankles. An extensive medical workup is inconclusive, and exploratory abdominal surgery is carried out which reveals a primary ovarian cancer with extensive spread to other abdominal organs. Her condition is judged to be too far advanced for surgical relief, and her life expectancy is estimated to be at most a few months. Despite her oft-repeated request to be told "exactly where I stand and what I face," Dr. E tells both the patient and her husband that the diagnosis is still unclear but that he will see her weekly as an out-patient. At the time of discharge she is feeling somewhat better than at admission, and Dr. E hopes that the family will have a few happy weeks together before her condition worsens and they must be told the truth.

Dr. E's behavior is clearly paternalistic: he has deceived Mrs. R for what he believes is her benefit; he knows that he is violating the moral rule against deception without her consent; and he views her as someone who is competent to give or refuse valid consent to being told other than the truth.

Dr. E could attempt to justify his parternalistic act by claiming that the evil—psychological suffering—he hoped to prevent by his deception is significantly greater than the evil, if any, he caused by lying. While this might be true in this particular case, it is by no means certain. By his deception, Dr. E is depriving Mrs. R and her family of the opportunity to make those plans that would enable her family to deal more adequately with her death. In the circumstances of this case as described, Mrs. R's desire to know the truth is a rational one; in fact, there is no evidence of any irrational behavior or desires on her part. This contrasts sharply to Mr. K's desire to kill himself, which is clearly irrational. Furthermore, although we regard Mr. K's desire not to be deprived of freedom, considered in isolation, as a rational one, we would not consider it rational for one to choose a high probability of death over the loss of a few days of freedom.

Nature of the justification procedure

In order to justify paternalistic behavior, it is necessary (not sufficient) that the evil prevented for S by the moral rule violation be so much greater than the evil, if any, caused to S by it, that it would be irrational for S not to choose having the rule violated with regard to himself. When this is not the case, as in Dr. T's deception of Mrs. R, then the paternalistic behavior cannot be justified. If it would not be irrational for S to choose suffering the evil rather than having the moral rule violated with regard to himself, then no rational person could universally allow the violation of the rule in these circumstances. This becomes clear when we specify what counts as "these circumstances." For us, specifying the circumstances is the same as specifying the kind of violation that one could publicly advocate. We hold that the only factors (or beliefs) that are relevant in specifying a given kind of violation are (or concern):

1. The moral rule(s) which is (are) violated
2. The probable amount of evil caused, avoided, or ameliorated by the moral rule violation (probable amount includes the kind and severity of the evil, the likelihood that the evil will occur, and the probable length of time it will be suffered)
3. The rational desires of the person(s) affected by the moral rule violation

Thus, these factors alone determine the circumstances we must take into account when we determine the kind of violation. What is then needed is to determine whether to publicly advocate this kind of violation. This is done by deciding whether the evil avoided or ameliorated by universally allowing this kind of violation outweighs the evil that would be caused by universally allowing it. If all rational persons would agree that the evil prevented by universally allowing the violation would be greater than the evil caused by universally allowing it, the violation is strongly justified; if none would, it is unjustified. If there is disagreement, we call it a weakly justified violation, and whether it should be allowed is a matter for decision.[3]

In order to make this point clearer let us consider the case of Mrs. R again. Dr. E is deceiving Mrs. R, and this requires moral justification. For us (as for almost all contemporary philosophers), morally justifying a violation of a moral rule requires that one be able to hold that everyone be allowed to violate the rule in the given circumstances. If one is violating a moral rule in circumstances in which one would not allow others to violate it, one is not acting impartially. And it is universally agreed that morality requires impartiality. Determining whether one would allow everyone to violate the moral rule in the given circumstances is best done by seeing if one would publicly advocate this kind of violation. Determining the kind of violation means determining the relevant circumstances in which the violation occurs. In the case under discussion, Dr. E is violating the rule against deception in circumstances in which there is a high probability that he is preventing mental suffering for several weeks; there is an equally high probability that he is depriving a person of the opportunity to make the most appropriate plans for her future; and the person affected by the deception has a rational desire not to be deceived.[4]

Given this description of the kind of violation, determining whether or not one would publicly advocate it is the same as determining whether one would allow everyone to deceive in the circumstances described. The following may make the point even clearer. Suppose someone ranks unpleasant feelings for several weeks as a greater evil than the evil involved in the loss of some opportunity to plan for the future. Should he be allowed to deceive those who may have a different rational ranking, if his deception will result in their suffering what the deceiver regards as the lesser evil? Would any rational person hold that everyone be allowed to deceive in these circumstances, that is, would any rational person publicly advocate this kind of violation? We think that once the case is made clear enough, no rational person would publicly advocate such a violation. For it amounts to allowing deception in order to impose one's own ranking of evils on others who have an alternative rational ranking. Allowing deception in such circumstances would clearly have the most disastrous consequences on one's trust in the words of others; thus, allowing this kind of violation would have far worse consequences on any rational ranking than not allowing it. Thus no rational person would publicly advocate such a violation. On this analysis, Dr. E's deception was an unjustified paternalistic act.

We are not maintaining that all lying to patients is unjustified. The following is a case of lying we regard as justified.

Case 8-3.

Mrs. V is in extremely critical condition after an automobile accident which has taken the life of one of her four children and severely injured another. Mrs. V is about to go into surgery and Dr. H believes that her very tenuous hold on life might be weakened by the shock of hearing of her children's conditions, so he decides to deceive her for a short period of time.

Anyone acting rationally who was presented with the option of being deceived for a short period of time or of greatly increasing his or her chance of dying would choose the former. Thus, if Mrs. V said that she wanted to know, knowing that the truth might kill her, we would not regard such a desire as rational. (We are excluding any other considerations, such as religious beliefs or her desire to die because of such a loss.) In these circumstances, we have deception which significantly decreases the chance of death and causes no significant evil. Would a rational person publicly advocate this kind of violation? What would be the effect of universally allowing this kind of violation? As far as we can see, there would be no significant loss of trust, and whatever loss might occur would seem to be more than balanced by the number of lives that would be saved. Thus, we hold that all rational persons would publicly advocate this kind of violation.

In discussing the justification of paternalism, it is very easy to fall into the error of supposing that all that we need do is compare evils prevented with evils caused and always decide in favor of the lesser evil. It is this kind of view, a relatively straightforward negative utilitarianism, which seems to be held by many doctors and may account for much of their paternalistic behavior. If A holds this view, then A thinks that if he is preventing more evil for S than he is causing S, he is justified in violating a moral rule with regard to S.[5] Taking this straightforward negative utilitarian view, one can easily understand why paternalistic acts of deception done in order to prevent or postpone mental suffering, as in the case of Mrs. R described above, are so common. Little if any evil seems to be caused, and mental suffering is prevented, at least for a time; thus it seems as if such acts are morally justified. Paternalistic acts by which someone is deprived of freedom, as in commitment, are recognized to be more difficult to justify, for then we have a significant evil—the loss of freedom—which is certain and which has to be balanced against the evil we *hope* to prevent.

But negative utilitarianism is not an adequate ethical theory. We do not regard the foreseeable consequences of a particular act as the only morally relevant considerations; they are, or course, very significant, for they help determine the kind of act involved. But if an act is a violation of a moral rule, we must consider the hypothetical consequences of universally allowing this kind of violation. In many cases, balancing the evil that would be caused against the evil that would be prevented in universally allowing this kind of moral rule violation will lead to the same moral judgment as if only the consequences of the particular act had been weighed. Probably for this reason some writers have taken the simpler balancing to be sufficient.

However, part of the standard philosophical literature against utilitarianism consists of examples in which it is inadequate to consider only the foreseeable consequences of a particular act. For example, consider a medical student who has not studied during the weeks preceding the state medical examinations he is now taking, but believes himself to be well qualified for the practice of medicine. He therefore cheats in order to increase his chance of qualifying for the practice of medicine and thereby prevent the unpleasant feelings to himself and his parents which would accompany failure. His cheating has not caused harm to anyone (assume that the exam is not graded on a curve) and on a simple negative utilitarian view would therefore be morally justified. We may therefore hypothesize that if we allow cheating for the purpose of decreasing unpleasant feelings, then some individuals, who may believe themselves qualified when in fact they are not, will cheat and thereby pass. Thus, we will destroy the value of these tests, on which we rely in part to determine who is qualified for medical practice. The effect of less qualified people becoming doctors is an increased risk of the population's suffering greater evils, such as pain and disability, and this outweighs the suffering caused to those who fail to qualify.[6] Thus, by using this more complex balancing, we conclude that such acts of cheating are not morally justified.

The procedure used in the earlier examples may be applied to the cheating example as follows: the amount of evil caused directly by this particular act of cheating is believed by the student to be nil; the evil avoided or prevented is unpleasant feelings for the student and his parents. There are supposedly no persons directly affected by this violation of the moral rule against cheating, so the rational desires of others need not enter the calculation. But suppose anyone who thought that no harm would come of his cheating, and that he could prevent a few people having unpleasant feelings, were allowed to cheat. We believe that the hypothetical long-range evils associated with universally allowing this kind of violation so far outweigh the unpleasant feelings that would be prevented, that allowing it would be irrational. Thus, it could not be publicly advocated.

Further examples of the justification procedure

Paternalistic acts committed by physicians may involve the violation of many moral rules, but the three most common violations seem to be depriving of freedom, deceiving, and causing pain or suffering. We believe that in all three violations there are both justified and unjustified paternalistic acts. We have given examples of depriving of freedom (Case 8-1, Mr. K) and of deception (Case 8-3, Mrs. V), which we have argued are justified, and an example of deception (Case 8-2, Mrs. R), which we have argued is not. Let us consider several more cases of medical paternalism in order to review the various components of our account of moral justification and to show how certain changes in these components can affect the final moral judgment reached.

Case 8-4.

Mr. L is a twenty-six-year-old single male patient with a past history of intense participation in physical activities and sports, who has suffered severe third-degree burns over two-thirds of his body. Both of his eyes are blinded due to corneal damage. His body is badly disfigured, and he is almost completely unable to move. For the past nine months he has undergone multiple surgical procedures (skin grafting, removal of his right eyeball, and amputation of the distal parts of the fingers on both hands). He has also required very painful daily bathings and bandage changings in order to prevent skin infections from developing over the burned areas of his body. The future he now looks forward to includes months or years of further painful treatment, many additional operations, and an existence as at least moderately crippled and mostly (or totally) blind. From the day of his accident, he has persistently stated that he does not want to live. He has been interviewed by a medical center psychiatrist and found to be bright, articulate, logical, and coherent. He is firm in his insistence that treatment be discontinued and that he be allowed to die. Nonetheless his physicians are continuing to treat him.[7]

According to our definition, Mr. L's doctors are acting paternalistically: they believe that saving Mr. L's life benefits him; they know they are causing him great physical and psychological pain without his consent; and they believe that he is competent to give or refuse valid consent to continued treatment.

Mr. L's physicians could claim that they are acting as they are because they believe that the pain they are causing him by continuing treatment is a lesser evil than the death that would occur should they stop. This is certainly a rational ranking on their part. If Mr. L agreed, then the physicians would still be violating a moral rule but with Mr. L's consent, and thus no moral dilemma would exist. But Mr. L ranks the evils differently: he prefers death to months of daily pain, months or years of multiple surgical procedures, and an existence as a deformed and crippled person. His ranking, like that of his physicians, is a rational one.

We would then say that the kind of violation being engaged in by the physicians involved their causing a great amount of pain by using their rational ranking of evils with regard to a person whose own rational ranking is different. No rational person would publicly advocate this kind of violation because of the terrible consequences of living in a world where great pain could be inflicted on persons against their rational desires whenever some other person could do so by appealing to some different rational ranking of evils of his own. Thus, when we universalize the kind of violation in which the physicians are engaged, we conclude that their paternalistic act is unjustified. (Note that the form and conclusion of this analysis are similar to those in the case of deceiving the cancer patient.)

We mentioned above that if Mr. L agreed with his physicians, no moral dilemma would exist. The case could also be varied in another way which would change our moral judgment. Suppose Mr. L had to undergo only one week of painful treatment and then had a high probability of resuming an essentially normal life. If he claimed to prefer death over one week of treatment, we would say that his ranking of evils was not rational. We would now describe the kind of violation being engaged in by the physicians as involving their preventing death by causing a great amount of pain for a short time on a person whose contrary ranking of evils is irrational.[8] A rational person could publicly advocate this kind of violation. Thus, universalizing this kind of violation, we conclude that the paternalistic intervention of the physicians would be justified.

The following two psychiatric cases involve physicians who wish to administer ECT without the patients' consent.

Case 8-5.

Mrs. D, a rather frail fifty-five-year-old woman, was admitted with a six-month history of a moderately severe depression. She was transferred from a community hospital where her local psychiatrist had tried two different antidepressant medications but had stopped both because of her marked drop in blood pressure when given even the usual low starting dosages.

When she came to the inpatient ward, Mrs. D was markedly depressed. She had lost a moderate amount of weight and was sleeping poorly but maintained a fairly adequate intake of food and water. Her psychiatrist recommended ECT to her. However, she firmly and consistently refused. She had had a close friend who had received ECT; while her friend's depression had improved at the time, she had killed herself a year later. Mrs. D acknowledged that ECT "may not have been responsible" for her friend's suicide but said she was terrified of it.

Her psychiatrist therefore devised a drug regimen in which she was given a very small nighttime dosage (10 mg) of desipramine, a tricyclic antidepressant, which was later increased slowly in small (10 mg) steps. He also gave her small morning and noontime doses of methylphenidate, a stimulant. She suffered from significant lowering of blood pressure, but with close nursing care this problem proved manageable. After two to three weeks, her depression began to respond. In another two weeks she was feeling quite well and her low blood pressure had ceased to be a significant problem.

Mrs. D's physicians viewed her as competent to give or refuse valid consent: she fully understood and appreciated the facts of her illness, knew that she was refusing her physicians' suggested treatment, and understood what her physicians believed might be the possible consequences of her refusal. Her refusal of ECT seemed, on balance, clearly mistaken if not irrational to her physicians. This was because ECT seemed to offer a high probability of significant and rapid benefit with extremely little risk, while it was not clear that continued drug treatment would help, at least not very quickly. It also seemed that her fear of ECT was based in part on what she herself seemed to acknowledge as the mistaken belief that ECT may have caused her former friend's death. Nonetheless, the harm she risked by not choosing ECT was limited to a prolongation of her marked depression. She was eating adequately, maintaining good fluid intake, and was not suicidal, so that there appeared to be no immediate danger of her dying. Therefore, the evils that ECT might ameliorate did not seem great enough to justify the evils that would be caused by coercively giving her ECT without her consent, especially since the alternative, continued and meticulous drug therapy, had at least some chance of success. Though we viewed Mrs. D's refusal of ECT as clearly mistaken we do not believe her psychiatrist would have been morally justified in paternalistically overriding.

Case 8-6.

Mrs. O, a sixty-nine-year-old woman, was admitted to the inpatient unit with a depressive illness of six months' duration. Approximately one year before admission, she was discovered on a routine examination elsewhere to have an enlarged spleen. No further studies were carried out.

Approximately six months before her hospitalization, Mrs. O's husband suffered a heart attack and was subsequently confined to a nursing home. She stated in retrospect that her "world went to pieces" at that time. She gradually became depressed and experienced characteristic changes in appetite, weight, and sleep. She refused to seek medical attention. Eventually her husband called his lawyer, who summoned the police to her home, where they found her in a state of neglect and brought her to the emergency room at her local hospital. She was admitted and noted to be depressed, but was alert, oriented, and cooperative. Positive physical findings included low red and white blood cell counts and a further increase in spleen size. She was seen by a consulting psychiatrist, who thought she was significantly depressed and recommended treatment with antidepressant medication. She agreed to take the medicine but did not improve. The patient's internist recommended a bone marrow examination and other laboratory studies. Mrs. O refused for reasons that she would not clearly discuss, saying only that she "didn't want to bother." She was transferred to a university medical center for a further attempt at evaluation and possible treatment.

This second evaluation confirmed the above impressions. She was seen by a neurologist and by a hematologist, who recommended a CT scan, an electroencephalogram, a spinal tap, and a bone marrow examination. The patient's clinical condition continued to deteriorate, and she began refusing most food and fluids. She refused to allow most of the recommended diagnostic tests to be performed. Repeated efforts by the staff and by her family to obtain her consent for these studies were unsuccessful and were now met by her saying that "I deserve to die."

It was felt she was indeed at risk of death through malnutrition and body chemistry disturbance resulting from inadequate fluids and nutrition. Her husband and son were informed of the seriousness of the situation. Her son obtained an attorney, went to court, and on the basis of the clinical details provided, obtained temporary legal guardianship of his mother. He then authorized proceeding with the diagnostic procedures deemed necessary. These were done despite her objections. The hematology consultant concluded that the most likely diagnosis was myelofibrosis. Her long-term prognosis from this disorder was thought to be questionable, but her prognosis for the next several years was quite good.

Her son authorized proceeding with ECT for her now severe melancholia. She was treated initially without her consent and over her stated objections.

After the second treatment, she gave verbal consent to further treatments; after the fourth treatment she became brighter in mood, began eating well, and was much more verbal. After a total of ten treatments, Mrs. O reported that she felt quite well. She exhibited a mild post-ECT confusion, which subsequently cleared. She was able to express appropriate feelings of sadness about her husband's illness. She said that she was very grateful that she had been treated. At her last follow-up visit, several months after discharge, she was doing quite well.

We view Mrs. O as having been competent to give or refuse valid consent. She understood and appreciated what her physicians believed to be the facts concerning her condition, knew that she was refusing her physicians' suggested treatment, and understood what her physicians believed would be the consequence of her refusal, namely, her death. Her refusal of ECT seemed seriously irrational to her physicians in that it seemed almost certain that she would die without ECT and there was little if any risk associated with it. She offered no reason, let alone an adequate one, for wanting to die, and this desire seemed to be associated with her depression, a kind of mental malady which is usually quite transient with adequate treatment. Thus, we believe that paternalistically forcing treatment on her (with her guardian-son's concurrence) was morally justified: that one can publically advocate such an intervention when the evil one is probably forestalling is so great, the evil one is perpetrating is so much less, and the patient's refusal is irrational.

All of the above cases involve the balancing of great evils such as dying, deception about terminal illness, and the infliction of severe pain. The paternalistic interventions described have been obvious and often dramatic: commitment to a mental hospital or lying about whether a person's child was alive. The health care professionals making the decisions were all physicians, and in most of the cases (though we have not mentioned this feature) the possibility of legal intervention was present, in the form of suits for negligence or battery or injunctions to stop treatment. However, we believe that the vast majority of paternalistic interventions in medicine take place on a smaller scale. The following case illustrates what we consider to be a much more common type of paternalism.

Case 8-7.

Mrs. J is a fifty-year-old patient in a rehabilitation ward who is recovering from the effects of a stroke. A major part of her treatment consists of daily visits to the physical therapy unit, where she is given repetitive exercises to increase the strength and mobility of her partially paralyzed left arm and leg. She is initially cooperative with Mr. Y, her physical therapist, but soon becomes bored with the monotony of the daily sessions and frustrated by her failure to adequately move her partially paralyzed limbs and by her very slow progress. She tells Mr. Y that she does not wish to attend the remaining three weeks of daily sessions. It is Mr. Y's experience that if patients like Mrs. J stop the sessions early, they do not receive the full therapeutic benefit possible and may suffer for the remainder of their lives from a significantly more disabled arm and leg than would be the case if they exercised now in this critical, early post-stroke period. Accordingly, he first tries to persuade her to continue exercising. When that is not effective, he becomes rather stern and scolds and chastises her for two days. She then relents and begins exercising again, but it is necessary for Mr. Y to chastise her almost daily to obtain her continued participation over the ensuing three weeks.

Mr. Y's scolding and chastising is paternalistic behavior by our account: he is causing Mrs. J some psychological pain and discomfort without her consent for what he believes to be her benefit. He further believes that Mrs. J is competent to give or refuse valid consent to physical therapy.

Mr. Y could attempt to justify his action by claiming that the relatively minor amount of evil he is inflicting by chastising her to exercise is much less than the relatively greater evil Mrs. J may experience by being significantly more disabled than necessary for the rest of her life. Thus, Mr. Y could claim that Mrs. J's ranking of the relevant evils is irrational. We would describe the kind of violation engaged in by Mr. Y as inflicting a mild degree of suffering on Mrs. J (through his chastising and her resumed exercising) by imposing his rational ranking of evils on Mrs. J, whose ranking is not rational. A rational person could publicly advocate this kind of violation, and we conclude that Mr. Y's paternalistic behavior is weakly justified.

Note that this case involves the balancing of evils which, while significant, are not of the intensity of our earlier cases. The amount of evil associated with the possibility of Mrs. J's needlessly greater lifelong disability seems just significant enough (and her balancing of the evils of three weeks of exercising versus greater lifelong disability seems just irrational enough) to weakly justify causing her a mild to moderate degree of transient suffering without her consent.

However, there are some kinds of violations we would think not justified in Mrs. J's case. We would think it unjustified to inflict intense physical pain on her to force her to exercise. The amount of evil associated with the possibility of her increased disability is not great enough so that one could publicly advocate that kind of violation.

Philosophically, the most interesting alternative to consider is the possibility of lying to her. Suppose Mr. Y told her that unless she continued to exercise for three more weeks she would regress and never be able to walk again. Suppose further that he knew that in her case this was untrue, that in fact she was almost certainly going to be able to walk in any event, but that, as described above, physical therapy would very likely decrease her ultimate level of disability. If Mr. Y did lie in this way, it might prove quite effective in quickly remotivating Mrs. J to exercise daily without the need for Mr. Y to chastise her at all. In fact, such a deception might cause Mrs. J less total suffering than daily chastising would if she now perceived the exercising as something she wanted to do because through doing it she was staving off the possibility of a certain inability to walk. Thus, using a simple negative utilitarian method of calculation, it might seem that if chastisement were justifiable paternalism, lying would be even more strongly justifiable. But when we describe the kind of violation that Mr. Y would commit by lying we see that this is not the case. It is true that Mrs. J's desire to discontinue physical therapy is, in the circumstances, one that we would regard as irrational. Thus we do not have lying in order to impose one person's rational ranking of the evils on another person's different rational ranking, but rather lying in order to substitute a rational ranking for an irrational one. Thus we must describe the kind of act as deception in order to prevent the possibility of a permanent (up to twenty to thirty years) though moderate amount of disability, and thereby causing a temporary (three weeks) mild physical discomfort (of physical therapy). Would a rational person allow lying in these circumstances, that is, would he publicly advocate this kind of violation of the rule against deception? We find this a difficult question to answer. Allowing lying in a situation where trust is extremely important (that is, between a doctor, nurse, or therapist and a patient) in order to prevent evil significant enough that it is irrational not to avoid it (that is, the possibility of permanent moderate disability) is an issue on which we think rational persons could disagree. The erosion of trust that would follow from universally allowing this kind of violation may have such significant evil consequences (for example, legitimate warnings may come to be disregarded) that it is not clear that even preventing a significant number of persons from suffering permanent moderate disability is enough to counterbalance these consequences. Our conclusion in this circumstance is that rational persons would disagree; some would publicly advocate violation, some would not.

The conclusion of the previous paragraph rests upon the assumptions that lying is the only method whereby one can get Mrs. J to continue her treatment and that lying will be successful in doing so. However, if we have an alternative to lying, namely, the method of chastising and scolding, then we hold that all rational persons would publicly advocate using this method rather than lying. We have found, in informal testing, that when presented with these alternative methods of getting Mrs. J to continue treatment, everyone regarded chastising and scolding as morally preferably to lying, and that some regarded lying as completely morally unacceptable. If one uses a simple negative utilitarian theory, this result cannot be accounted for since in this particular case lying results in no more and almost certainly less overall suffering than chastising and scolding. However, if one applies the method of justification that we have been presenting, these moral intuitions are accounted for quite easily. For using *only* the relevant facts of the particular case—(1) the moral rules violated: causing pain (the unpleasantness caused by the scolding) versus deception (lying about consequences of stopping treatment); (2) the evil probably prevented: permanent moderate disability in both cases; and (3) the same rational desires in both cases—as determining the kind of violation and then seeing whether one could publicly advocate this kind of violation, one will end up with the result that accords with the moral intuitions that one actually finds.

We believe the above case is typical of a multitude of everyday situations in medicine in which doctors, nurses, and other health care workers act or are tempted to act paternalistically toward patients. Consider the problems presented by the patient with emphysema who continues to smoke, by the alcoholic with liver damage who refuses to enter any treatment program, or by the diabetic or hypertensive patient who exacerbates his disease by paying little heed to dietary precautions. Each of these patients is apt to stimulate paternalistic acts by a variety of health care professionals (as well as members of his own family). We think it is important to determine, whenever possible, which if any kinds of paternalistic acts are justifiable and which are not.

Another theory of justification

As noted earlier in this chapter, several writers have held, as we do, that some paternalistic acts are justified and others are not. It would be out of place in this book to consider these other theories exhaustively, but one approach to justification that physicians sometimes use deserves discussion: the "thank-you" test. According to this test, one may justifiably act paternalistically if one is certain that at some later time one will be thanked by the person (patient) toward whom one is acting. Alan Stone's (1975, p. 70) "thank you theory of civil commitment" is a variation of this approach. John Rawls (1971, p. 249) has similarly written: "We must be able to argue that with the development or the recovery of his rational powers the individual in question will accept our decision on his behalf and agree with us that we did the best thing for him." James F. Childress (1979, p. 26) has recently labeled these approaches "ratification theories," though it is not clear whether he wholly agrees with them.

These theories seem simple and therefore attractive. However, while the recipients of justified paternalistic acts are often subsequently grateful, it is important to see that "thank-you" theories are inadequate accounts of the justification process.

The first criticism of these theories is that obtaining future thanks or ratification is neither necessary nor sufficient to justify paternalistic acts. To see that it is not necessary, consider the following cases. First, imagine a case in which one is certain a patient will be thankful (and the paternalistic act thus justifiable according to this test) and then imagine that this particular patient, a rather grudging person, later is not thankful (which could certainly happen in a minority of cases, as with patients with personality disorders). One would not want to say that the paternalistic action had turned out to be unjustified after all. A theory of justification should depend on something less capricious than a given patient's eventually revealed character style. A second illustrative case: you have deceived a patient, and it would be counterproductive to reveal the deception, so that patient never knows that you have acted paternalistically toward him. Here, one is tempted to say that the patient would thank you if it were possible to let him know what had happened. This makes it even clearer that what is important is not the patient's actual thanks but rather that the justification depends on factors known at the time of the decision to act paternalistically.

Moreover, receiving future thanks is not sufficient for justification either. Childress notes one reason: "some interventions, such as brain-washing, may create the conditions of their own ratification" (p. 26). But it is even more generally true that some patients, for interesting psychodynamic reasons, are so obeisant toward physicians that they will forgive them and even thank them for a variety of what appear to be unwarranted violations of the moral rules with regard to them. One would not want that kind of patient obeisance to be sufficient justification for particular paternalistic actions either.

The second criticism of thank-you theories is more basic: one must ask where the judgments about whether one will later be thanked come from. They must come from certain features of a given case that exist at the time that one decides to act paternalistically. What are these features? We believe that they are just the features we use in deciding whether or not a paternalistic act is justified: (1) the moral rule(s) to be violated, (2) the probable amounts of evil to be caused, avoided, or ameliorated, and (3) the rational desires of the person toward whom one is acting.

Thus, thank-you theories have only an illusory simplicity and do not allow one to avoid the task of isolating the criteria for justified paternalism.[9] We believe that the plausibility of these theories comes from the fact that we usually judge whether a patient will say "thank you," at least implicitly, by using the criteria we have developed in our theory of justification.

Notes

1. This chapter is adapted from Gert and Culver (1979).
2. In this chapter we apply the general theory of moral justification contained in Gert (1975).
3. We think that following the procedure outlined in this chapter will reduce the number of cases in which there is disagreement. In those cases of disagreement that remain, we believe that some type of decision-making body should be consulted whenever possible. This is an important practical matter that we do not discuss further here.
4. We realize that this description lacks precision, but we believe that often no greater precision is possible. This is one factor that makes coming to decisions sometimes difficult. See Aristotle, *Nicomachean Ethics*, 1.3.1094b.
5. Since it is only the evils caused and prevented with regard to S that are relevant, we are not involved in the kinds of interpersonal comparisons of utility that seem to present so many problems to utilitarian theorists.
6. It is this fallibility of persons, their inability to know all the consequences of their actions, which explains not only why no rational person would publicly advocate cheating simply on the grounds that no one would be hurt by it but also why moral rules are even needed. We cannot consider the nature of a violation after we see how things actually turn out but must do so when the violation is being contemplated; then the fallibility of persons will play its proper role. See Gert (1975, p. 126).
7. This case is adapted from Case No. 228, "A Demand to Die" in *Hastings Center Report*, 1975, *5*, 9.
8. We realize that there will not be universal agreement on the rationality of all examples of the rankings of evils, but we think there will be agreement on many, and thereby some of the initially unclear cases will become much clearer (see notes 3 and 4 above).

9. Childress acknowledges this point in a footnote (p. 26): "Since many individuals who are subject to paternalistic interventions in health care will never regain rational powers, the ratification theory often takes a hypothetical form: what individuals *would* consent to if they *could* consent. This version of the ratification theory, of course, appeals to some vision of what rational individuals do and should desire."

References

Beauchamp, Tom L. Paternalism and bio-behavioral control. *The Monist*, 1977, *60*, 62–80.

Childress, James F. Paternalism and health care. In Wade L. Robison and Michael S. Pritchard (eds.), *Medical Responsibility*. Clifton, N.J.: Humana, 1979, pp. 15–27.

Gert, Bernard. *The Moral Rules*. New York: Harper Torchbooks, 2nd paperback ed., 1975.

Gert, Bernard, and Culver, Charles M. The justification of paternalism. *Ethics*, 1979, *89*, 199–210.

Rawls, John. *A Theory of Justice*. Cambridge: Harvard Univ. Press, 1971.

Stone, Alan A. *Mental Health and Law: A System in Transition*. Rockville, Md.: U.S. Department of Health, Education, and Welfare, 1975.

9

The Morality of Involuntary Hospitalization

Introduction

There is a good deal of recent dispute both inside and outside of psychiatry about the procedure of involuntary hospitalization.[1] (We use "involuntary hospitalization" to cover both two-to-three-day "emergency" detentions and longer civil commitments. This chapter deals mainly with detention but refers to commitment when appropriate.) While writers such as Szasz (1978) believe that involuntary hospitalization should be eliminated entirely, it is more common to find disagreement about how wide or narrow the grounds for detaining patients should be.

Those who advocate narrow grounds are currently in the ascendancy. They prefer that detention be limited to those patients who not only are mentally ill but who also exhibit evidence of clear and immediate physical dangerousness to themselves and/or others. Some commentators have suggested adding further conditions, such as "incompetence of the patient to make treatment decisions" and evidence that the patient's condition is in fact treatable (see American Bar Association [1977] for a well-documented presentation of a narrow-grounds position).

Those who urge wider grounds also include mental illness as a necessary condition but wish to eliminate physical dangerousness as a necessary condition, substituting instead any severe disruption of personal functioning (see Chodoff [1976] for a well-argued presentation of this point of view; Chodoff includes "treatability" as a third necessary condition).

Those who advocate wider grounds often point out that physical dangerousness is but one of many possible manifestations of severe mental illness and that it seems cruel and inconsistent not to treat other severely disabled patients who seem urgently to require it. Narrow-grounds advocates, however, feel that any relaxation of criteria in the direction of a "disruption of functioning" rule would give physicians such wide discretionary power that civil liberties would generally be threatened.

We want to offer here a conceptual and moral analysis of the detention procedure which we believe leads to some clarification of the issues under dispute. First, we shall consider paternalistic detention. We believe that when one appreciates that most detention is paternalistic and then applies the justification process described in Chapter 8 to it, one can see more clearly the sources of some of the current disputes in this area. Next, we will briefly consider some problems raised by nonpaternalistic detentions as well as a number of other critical issues.

Paternalistic detention

Most detentions are paternalistic acts, according to our definition. They are done with the belief that they are for the patient's benefit; psychiatrists realize that the detention involves breaking the moral rule against deprivation of freedom; and they know that they do not have the patient's consent.

Psychiatrists also believe that the patient is competent to give or refuse at least simple consent to be hospitalized. The majority of detained patients, in fact, are competent to give or refuse valid consent, as defined in Chapter 3: they know and understand that they have a mental malady (usually depression), know that their physicians believe they require treatment in an in-patient setting for their own protection, and realize that they are refusing to consent to such treatment and are therefore being detained without their consent. But a minority of detained patients, while they are competent to give simple consent, are not competent to give valid consent. For example, some manic patients who are involved in reckless and potentially self-injurious hyperactive behavior may rigidly deny that there is anything wrong with them. Thus, they may refuse treatment or hospitalization without any understanding that they do, in fact, have a malady. On our account, they are competent to refuse simple consent for hospitalization but not competent to refuse valid consent. Since they are competent to refuse simple consent, they satisfy the fourth feature of our definition of paternalistic behavior, and detaining them would be a paternalistic act if the definition's first three features were also satisfied.

However, a patient who is incompetent to give or refuse simple consent would not satisfy the definition's fourth feature, and hospitalizing such a patient would not be paternalistic. For example, a patient in a catatonic stupor might be hospitalized with his next-of-kin's consent. Such a hospitalization can be regarded as involuntary, but it is not paternalistic.

The justification of paternalistic detention

Since detention always involves violating a moral rule with respect to a patient, it always requires moral justification. In order for paternalistic detention to be justified according to the procedure outlined in Chapter 8, (1) the amount of evil probably prevented or ameliorated for the patient by detaining him must be very great; (2) the amount of evil perpetrated on the patient by detaining him must be so much less than the evil probably prevented that it would be irrational for him to prefer the latter to the former; and (3) the patient must have no adequate reason, for example religious or political, for suffering the evil that detention is intended to prevent. It is only when these conditions hold that all rational persons could consider universally allowing this kind of moral rule violation to occur.

Let us illustrate this justification procedure by applying it to the case described in Chapters 2 and 8, where detention intuitively seemed to be clearly justified:

Case 9-1.

Mr. K was brought to the emergency room by his wife and a police officer. Mrs. K had confessed to her husband earlier that evening that she was having an affair with one of his colleagues. He had become acutely agitated and depressed and, after several hours of mounting tension, told her he was going to kill himself so "you'll have the freedom to have all the lovers you want." She became frightened and called the police because there were loaded guns in the house and she knew her husband was an impulsive man.

In the emergency room Mr. K would do little more than glower at Dr. T, his wife and the officer. He seemed extremely tense and agitated. Dr. T decided that for Mr. K's own protection he should be hospitalized but Mr. K refused. Dr. T therefore committed Mr. K to the hospital for a seventy-two hour emergency detention.

To summarize the justification procedure, the *kind of violation* which is involved in this case consists of A's temporarily (seventy-two hours) depriving S of his freedom and thereby also causing S a mild to moderate degree of psychological suffering (some of it temporary) in order possibly to prevent a very great permanent evil (death or serious injury); this is occurring in a setting where S apparently has no adequate reason for wishing to die and where for S to prefer the serious risk of death over the less serious evils associated with hospitalization seems irrational. It is important to note that there is a high probability that much of the evil imposed by detention will be temporary because Dr. T believes Mr. K's condition is rather quickly reversible. If the evil had to be imposed indefinitely, the situation would be altered significantly. We believe that essentially all rational people would agree that this kind of violation should be universally allowed and thus that Dr. T's paternalistic action is (strongly) justified.

According to our justification procedure, detention of apparently suicidal patients is morally justified only when (1) failure to detain would, with sufficiently high likelihood, be followed by death or very serious injury, (2) the person has no adequate reason for suicide, and (3) detention probably will, in a relatively short time, lead to a significant decrease in the likelihood of serious self-injury. It is only in these instances that a rational person could universally allow the imposition of less serious and generally temporary evils.

This is a very stringent test. Involuntary hospitalization would never be allowed *just* because someone was "mentally ill," since the test demands the specification of what evils of what magnitude are probably being prevented or ameliorated; the (mere) designation of someone as mentally ill is imprecise on this point, though the presence of mental illness does play a role in the justification of detention (see below). It is true that many patients who have mental maladies are suffering evils of one kind or another for which they have no rational desire. But for detention to be warranted, it is necessary that the present or probable future evils be very great *and* that there be a reasonably high likelihood that hospitalization will ameliorate or prevent them. That is, it must be irrational to prefer the possible evil to be suffered without hospitalization to the certain evil of hospitalization.

Issues in the justification of detention

The seriousness of the evils prevented and their probability of occurrence

How are we to judge which present or possible future evils are sufficiently serious to warrant detention? The seriousness of an evil is determined by several factors, among which is the impact it will make on the person's life at a given time. On this test, death is clearly the most serious evil, for nothing has greater impact than death. Thus, potential suicide (of sufficient likelihood) is almost universally taken as grounds for detention. On this life-impact test, serious bodily injury, great physical or psychological pain, and substantial social or economic losses can be considered as relatively equal. A second factor is how permanently the harm will affect the rest of the person's life. Here again, nothing is more permanent than death. But of the harms considered more or less equal on the previous test, bodily injury will usually have a much more permanent effect than any of the others; for example, brain damage and blindness are permanent. On the other hand, it is almost always possible to reverse social and economic losses and to recover from psychological or even physical pain.

We think that only the evils of death and serious bodily injury, which have both great initial impact and permanence, are sufficient to make it irrational for one not to prefer experiencing the evil of detention. This limitation gains added support when we recognize that the evils of detention are certain but the evils to be prevented are only possible. We know that by detaining we deprive a person of freedom, but it is almost never *very* certain that the person will suffer an evil if he or she is not detained. The fallibility of psychiatrists (or anyone else) in predicting this kind of very infrequent behavior is well known (e.g., see Livermore, Malmquist, and Meehl, 1968).

Given the certainty of causing evil by detention, the evils to be prevented must not only be great, there must also be a significant probability of their occurring in order to justify paternalistic detention. Concentration on particular cases, especially after the fact, that is, when failure to detain *has* been followed by suicide or another extremely serious consequence, tends to lead one to advocate more lenient standards for detention. In this controversy, we are clearly against using what happens after the fact in a *particular* case as a legitimate argument for more lenient standards for detention. Only the information available at the time of possible detention can count in determining whether detention is justifiable or not. Consider the following example offered by Chodoff (1976):

Case 9-2.

Passersby in a campus area observe two young women standing together, staring at each other, for over an hour. Their behavior attracts attention, and eventually the police take the pair to a nearby precinct station for questioning. They refuse to answer questions and sit mutely, staring into space. The police request some type of psychiatric examination but are informed by the city attorney's office that state law (Michigan) allows persons to be held for observation only if they appear obviously dangerous to themselves or others. In this case, since the women do not seem homicidal or suicidal, they do not qualify for observation and are released.

Less than thirty hours later, the two women are found on the floor of their campus apartment, screaming and writhing in pain with their clothes ablaze from a self-made pyre. One woman recovers; the other dies. (There is no conclusive evidence that drugs were involved.)

Chodoff seems to offer this case to support wider grounds for detention but we think it is clear, on the facts given, that detention would not have been morally justified. The correlation between the kind of behavior described and subsequent suicide would be too low. The case is a tragedy, but there will always be tragic cases regardless of how wide one's criteria are for detention: there will inevitably be people who fall just outside the criteria who subsequently kill themselves.

The likelihood of preventing harm to the patient

How likely is it that psychiatric detention does prevent serious harm to a person? Unfortunately, it seems true that there are few data available to answer this important question.

If we knew that 90 percent of those who were detained (we are considering only those without an adequate reason) would have killed or seriously injured themselves without detention and that these evils are prevented for an appreciable period of time with, say, seventy-two hours of detention, then one element of the justification process would be strongly satisfied; that is, it would be irrational to prefer a 90 percent chance of death or serious injury to seventy-two hours of detention. By contrast, if only 1 in 10,000 would kill or injure themselves without detention this criterion would not be satisfied, for then the probable amount of evil prevented is extremely small and it would no longer be irrational to prefer this small probability of death over the evils of detention.

Unfortunately, the actual percentage is at present unknown. Available studies (American Bar Association [1977], p. 87) suggest it is in the range of one out of six if detention is appropriately limited to persons who show evidence of certain high-risk attributes (marked depression, recent history of impulsive and/or suicidal behavior, etc.).

Assume for a moment that the actual figure for a defined subclass of suicidal persons is 16 percent. In order to assess the rationality of preferring a 16 percent chance of death to three days of detention without the distorting feature of suicide, let us change the case somewhat. Suppose a man had just suffered a heart attack and it was known that his risk of death was essentially zero if he remained in bed for one to three days (whether he liked it or not) but was 16 percent if he insisted, without any adequate reason, on being up and about. We believe that nearly everyone would view it as irrational if he did insist on getting up, even though we knew that by universally recommending bedrest we were needlessly confining five people for every one person for whom it would be necessary.

Suppose the figure were 10 percent. We believe that the large majority of people would still view it as irrational not to choose to be confined for a few days. How low the percentage would have to go before a significant number of people would think it rational to get out of bed we do not know. However, it seems very likely that if our subclass of suicidal persons were defined stringently enough, then the percentage of them who would suffer serious evils without brief detention would be sufficiently high that most people would view it as irrational to prefer that risk to the evils associated with brief detention. For example, suppose we detained only those who were markedly depressed, had a history of impulsive behavior, and were apprehended when they were apparently about to kill themselves. (We will assume that there is some subgroup in which detention does prevent harm with sufficient likelihood that it would be irrational for someone in this subgroup not to prefer it, but more empirical research on this question is badly needed.)

The rationality of desiring the evils prevented

We are not justified in detaining someone to prevent his suffering an evil if it does not appear irrational for the person to want to suffer that evil. If the person has an adequate reason for suffering a serious evil, then detention is unjustified. Thus we should not detain a pain-ridden, terminal cancer patient who has only weeks to live if we learn that the patient intends to take a fatal dose of barbiturates: for such a person to prefer death to the evils of weeks of agony followed by certain death seems a rational choice. It would also be rational, of course, to choose to live out one's life under these circumstances rather than end it prematurely. But to allow a physician to detain a patient making the former choice would be (universalizing according to our justification procedure) to allow physicians to substitute forcibly *their* rational rankings of evils for their patients' equally rational rankings, and that would be a *kind of violation* of a moral rule that would be unacceptable to rational persons.

We have noted that while it sometimes seems relatively clear that there is no adequate reason for persons to desire an evil, in other cases it is harder to judge. It is usually easier to judge the rationality of desires involving physical harm; for it is often clear whether or not the person has an adequate reason, for example, the above terminal cancer patient for wanting to die. However, the rationality of what appear to be desires for social, economic, or psychological harm is more difficult to judge. It is often not clear that, say, a man's squandering his money or acting in a way that will ruin his reputation is clearly irrational. This is another reason, in addition to lack of permanence, why we may be more reluctant to detain someone to prevent apparently irrational nonphysical harms: the greater likelihood that we may be wrong. For example, one patient with a history of mania ushered in a further episode by spending large sums of money on what appeared and later proved to be worthless speculative investments, to his wife's (and later his own) great distress. However, it seems unlikely that speculative investment would ever be so clearly irrational as to justify detention or other restriction of freedom.

The following case is more difficult:

Case 9-3.

Mr. D is a thirty-year-old single author who suffered an episode of mania two years ago. Since then he has been taking lithium and apparently doing well. He has just used up the small savings that had enabled him to continue writing, when he sells the novel he has been working on for the past several years. After the sale he abruptly stops taking lithium and within two months develops signs of mild to moderate hyperactivity, pressured speech, anorexia, and insomnia. Soon thereafter, a good friend of his becomes aware that Mr. D has accumulated in his apartment, in cash, the proceeds from the sale of his book as well as the small amount of money remaining in his savings account. These funds constitute Mr. D's sole means of support for the foreseeable future; the friend is astounded to learn that Mr. D, normally a careful and frugal man, has begun burning the money. Mr. D gives his friend no coherent explanation for doing so but tells him a rambling and jumbled story which involves not trusting the bank and also fearing that his home may be robbed. He wishes to destroy the money rather than let it fall into someone's hands. Mr. D's friend is alarmed by Mr. D's behavior; with some effort he persuades Mr. D to accompany him to see Mr. D's psychiatrist, Dr. L. After Dr. L interviews Mr. D and becomes familiar with the situation, he informs Mr. D that he ought to enter the hospital or at least begin taking his lithium again; meanwhile, he should let someone else safeguard his money.

Suppose Mr. D will accept none of Dr. L's suggestions. Should he be hospitalized involuntarily?

We think not, though this is a difficult case. However, we believe the evils Mr. D is experiencing are not sufficiently intense or permanent in their effects to justify imposing the evils of detention. Also, while Mr. D's reasons for burning his money do not seem adequate, it is also true that socially aberrant actions like money burning are sometimes carried out for quite adequate, if unusual, reasons; thus, we hesitate to allow physicians to detain patients for such actions because of the definite possibility that physicians may incorrectly assess the rationality of the act. Even if one were convinced, after knowing the many facts in this case, that Mr. D should be detained, it would be difficult to write a statute that would be sufficiently detailed to cover this particular kind of case. Any statute of ordinary detail that would include Mr. D would also include other cases for which detention would be inappropriate.

Of course, Mr. D's mania may progress to a point where his hyperactivity, loss of appetite, and insomnia do become life-threatening; then detention might be justified. But it might not progress to that point. It is his mania, in fact, which makes this a difficult case—a point to which we will soon return.

The evils imposed and the role of mental maladies

For detention to be justified, the evils imposed by detention must almost always have a high likelihood of being transient. This is where the role of mental maladies in justifying detention becomes important. For we know as a matter of empirical fact that those irrational desires which accompany recognizable conditions of mental malady (e.g., mania, depression, and schizophrenia) usually disappear when the mental malady is treated. So, there is a high probability that the evils we impose need be only transient, thus less in total amount, thus easier to justify imposing.

For how long should detention be allowed? If the evils associated with an episode of mental malady prove not to be readily reversible with a brief detention (two to three days), what should be done next? It would be of practical utility to have available an additional two- to three-week extension of detention if there were compelling evidence that the patient would probably suffer severe evils if released. Because such an extension would greatly increase the amount of evil imposed on the patient, it should not be allowed unless the patient has recourse to due process in the setting of a court or some impartial body; this would then represent a civil commitment. An additional issue which should be considered at the same time is whether treatment should be given involuntarily. One justification for involuntary treatment would be that the patient, even though in the hospital, would likely die or suffer serious injury without it. If involuntary treatment were mandated, it would usually be pharmacologic, and two to three weeks would usually be sufficient for the drugs to have a therapeutic effect.

Only in extremely rare cases would commitment for a longer period of time be morally justified. It would be necessary (but not sufficient) to show that the evils associated with prolonged confinement were significantly less than the evils associated with freedom and that there was a significantly high probability that the patient could be successfully treated with prolonged hospitalization. Usually such patients should not be committed because the total amount of evil imposed by a long commitment is too great to be justified (see Chodoff [1976], p. 499, and Livermore, Malmquist, and Meehl [1968]).

We agree with Stone (1975) that in many cases it might be critical whether treatment were available in the hospital to which detention is made: as noted above, it is unjustifiable to impose the evils of detention when there is little likelihood of reducing the evils experienced by the patient. We are unclear, however, how much force this point has, since in most cases temporary confinement alone eliminates the risk of immediate self-harm and substantially reduces the risk of future self-harm (see American Bar Association [1977], p. 87). Of course, if a longer commitment were necessary, then treatment availability might be critical.

The problem of recurrent manic episodes

The decision not to intervene in Case 9-3 is not an easy one. Many manic patients like Mr. D manifest recurrent cyclical behavior which (1) can be financially or psychologically (but not physically) very harmful, (2) seems, by its conjunction with other known symptoms, to be almost certainly part of a manic episode, and (3) would almost certainly be treatable (reversible) in a short period of time. It appears that manic patients are responsible for a large number of the troublesome instances in which physicians feel some urge to detain but are unable to do so when narrow guidelines prevail. The situation is made more poignant by the fact that some patients subject to recurrent manic attacks are themselves concerned and apprehensive about the harm they may do if and when future episodes develop. Would it be possible to allow such patients to appoint a proxy or "guardianship committee" (perhaps composed of the spouse or parent, patient-designated attorney, and psychiatrist) which, if in unanimous agreement, could sanction future detention and treatment decisions, using broader criteria than immediate physical harmfulness? This might be called an Odysseus Pact or Odysseus Transfer and be thought of as a psychiatric analogue of the "living will." Such a committee might, in particular cases, choose to confiscate a person's funds temporarily rather than deprive him or her of freedom. We can see some difficulties with such a procedure but believe it merits discussion. Suppose a patient were insistent that he be allowed to appoint such a committee, and preferred to take what he believed to be the very slim chance of unjustified detention (or confiscation of funds) against the greater likelihood of future psychological and/or economic harm. Are there moral grounds on which one could refuse his request?

Harmfulness toward others

It seems clear that detaining a man to prevent him from inflicting harm on himself (without an adequate reason) may benefit him and is sometimes justifiable; what about detaining a man to prevent him from harming others? If a man is detained *primarily* because of potential harm to others, then clearly we are not dealing with paternalistic detention. Can detention of the mentally ill sometimes be justified on nonpaternalistic grounds? We believe it can. But it should be emphasized that an intervention to prevent someone from harming himself and an intervention to prevent someone from harming others are, from the standpoint of moral justification, two entirely different kinds of actions. They are commonly compounded in discussions on this subject by the phrase "danger to self or others," but this breeds confusion.

The moral situation created when one detains individuals who are potentially harmful to others is one in which the physician deprives S of freedom in order to prevent S from inflicting serious evils on T. Several features of this situation distinguish it from the situation in which S is potentially harmful to himself. The degree of irrationality of S's potential act plays almost no part in justifying this situation for one person's desire to harm another is never in and of itself irrational. Its irrationality depends upon the evil consequences that S might experience subsequent to his harming others (e.g., punishment); these consequences are usually much less certain than in acts of self-harm. Often it would not be irrational to prefer the risk of these consequences to the evils of detention. Thus, the kinds of tests we believe useful in assessing the justification of paternalistic interventions do not serve us here.

Should we then conclude that those individuals who are potentially harmful to others are to be left solely to the police? This would be a practical and simple solution to the problem, yet it runs counter to our intuition in many cases. For example, consider two men who are brought to the emergency room by the police. In each instance, the police have been called because the man's wife had just reported that her husband has threatened her life. Each man admits to the emergency room psychiatrist that this is true.

In Case 9-4 the man has a history of paranoid schizophrenic episodes and as a part of a developing delusional system has recently heard voices instructing him to kill his wife. The psychiatrist feels there is a significantly high probability of the man's harming his wife if he returns home.

In Case 9-5 there are no symptoms of any major mental illness. There is a background of chronic marital discord, and the psychiatrist feels there is a significantly high probability of the man's harming his wife if he returns home.

Many believe that we are justified in detaining the man in Case 9-4 but not in Case 9-5. Of course, a psychiatrist might be *motivated* to detain both men simply out of a desire to protect their wives, but potential harmfulness to others does not provide a sufficient *justification* for *psychiatric* detention.

In general, we believe that psychiatrists should intervene in those cases in which the patient is seen as not being fully responsible for carrying out his harmful action. Were the man in Case 9-5 to harm his wife, he would be seen as fully responsible (from the information given), while the man in Case 9-4 would not. To use the terminology developed in Chapter 6, the man in Case 9-5 appears to have the volitional ability to will or to refrain from willing to harm his wife, while the man in Case 9-4, because of his mental malady, may not have the volitional ability to will not to harm her. We will not develop his matter further here but simply summarize by saying that it seems to be the *irrationality* of the potential act which justifies intervention with respect to self-harm and the *unvoluntariness* which sometimes justifies intervention with respect to harm to others.

It is true that psychiatrists are often under pressure (sometimes from the police themselves) to detain persons in cases like Case 9-5, because the man *does* appear harmful and there is a feeling that something should be done. However, we believe that if the psychiatrist determines that there is no mental malady present that would excuse the criminal act, then the person should be managed through police procedures: for example, using assault laws if the person has threatened others and/or issuing court orders temporarily prohibiting the individual from seeing a person he has been threatening.

Conclusion

Psychiatric detention is almost always a paternalistic action, and although it always requires moral justification, in some cases justification can clearly be provided.

We advocate relatively narrow grounds for psychiatric detention. We believe that one necessary condition for the justification of paternalistic detention is that the evils likely to be prevented for the patient by detaining him are of such great life impact and permanence that it would be irrational for the patient to prefer them to the evils associated with being temporarily detained. A second necessary condition is that the person have no adequate reason for suffering the evils that detention is intended to prevent. The third necessary feature is that there be both a significant probability the evils will occur without detention and that this probability be significantly diminished by detention. We think that these are the important features of cases that account for the moral judgments we make about particular acts of paternalistic detention. When all of these conditions are present, it seems not only that the physician is justified in detaining the person but also that he would be failing in his duty as a doctor if he did not do so.

The dispute about what grounds are adequate for paternalistic detention seems not only inevitable but welcome. In most acts of paternalistic detention there is a judgment by A that S's benefit justifies depriving S of his freedom; that judgment is, of course, fallible. We are sensitive to evil intentions masquerading as good; we should be equally aware that purely beneficent paternalism can be distinctly immoral. In a fine essay, Ira Glasser (1978) has written: "The assumption of benevolence must be seen as an insufficient reason to grant unlimited discretionary power to service professionals. We must begin, at least legally, to mistrust service professionals as well as depend on them, much as we do the police." By being as clear as we can be about what criteria justify detention, we can hope to reduce the unnecessary evil committed by beneficent but fallible psychiatrists. But some acts of commitment seem to be not only morally justified but morally required. Thus, we should also mistrust those civil libertarians who, out of their own benevolent intentions, urge us to regard temporary deprivation of freedom as a harm outranking even a significant possibility of death or serious injury for the patient. That policy too could lead to unnecessary evil for patients. The dispute is welcome because each side tends to overlook the fact that it can commit excesses by acting with sincerely benevolent motives. We believe that what is best for patients has the best chance of emerging when both sides to the dispute argue their case in the clearest possible fashion.

Notes

1. This chapter is adapted from Culver and Gert (1981).

References

American Bar Association. *Mental Disability Law Reporter*, 1977, *2*, 73–159.

Chodoff, Paul. The case for involuntary hospitalization of the mentally ill. *American Journal of Psychiatry*, 1976, *133*, 496–501.

Culver, Charles M., and Gert, Bernard. The morality of involuntary hospitalization. In Stuart F. Spicker, Joseph M. Healcy, Jr., and H. Tristram Englehardt, Jr. (eds.), *The Law-Medicine Relation: A Philosophical Exploration.* Boston: Reidel, 1981, pp. 159–175.

Glasser, Ira. Prisoners of benevolence: Power versus liberty in the welfare state. In Willard Gaylin, Ira Glasser, Steven Marcus, and David Rothman, (eds.), *Doing Good: The Limits of Benevolence.* New York: Pantheon, 1978, pp. 97–168.

Livermore, Joseph M., Malmquist, Carl P., and Meehl, Paul E. On the justification for civil commitment. *University of Pennsylvania Law Review*, 1968, *117*, 75–96.

Stone, Alan A. Comment on M. A. Peszke: Is dangerousness an issue for physicians in emergency commitment? *American Journal of Psychiatry*, 1975, *132*, 825–828.

Szasz, Thomas S. Involuntary mental hospitalization: A crime against humanity. In Tom L. Beauchamp and LeRoy Walters (eds.), *Contemporary Issues in Bioethics*. Encino, Cal.: Dickenson, 1978, pp. 551–557.

10

The Definition and Criterion of Death

Much of the confusion arising from the current brain death controversy is due to the failure to distinguish three distinct elements: (1) the definition of death; (2) the medical criterion for determining that death has occurred; and (3) the tests to prove that the criterion has been satisfied. We shall first define death in a way which makes its ordinary meaning explicit, then provide a criterion of death which fulfills this definition, and finally, indicate which tests have demonstrated perfect validity in determining that the criterion of death is satisfied.[1]

The definitions of death which appear in legal dictionaries and the new statutory definitions of death do not say what the layman actually means by death but merely set out the criteria by which physicians legally determine when death has occurred. *Death*, however, is not a technical term but a common term in everyday use. We believe that a proper understanding of the ordinary meaning of this word or concept must be developed before a medical criterion is chosen. We must decide what is ordinarily meant by death before physicians can decide how to measure it.

Agreement on the definition and criterion of death is literally a life-and-death matter. Whether a spontaneously breathing patient in a chronic vegetative state is classified as dead or alive depends on our understanding of the definition of death. Even given the definition, the status of a patient with a totally and permanently nonfunctioning brain who is being maintained on a ventilator depends on the criterion of death employed.

Defining death is primarily a philosophical task; providing the criterion of death is primarily medical; and choosing the tests to prove that the criterion is satisfied is solely a medical matter.

The definition of death

Death as a process or an event

It has been claimed that death is a process rather than an event (Morison, 1971). This claim is supported by the fact that a standard series of degenerative and destructive changes occurs in the tissues of an organism, usually following but sometimes preceding the irreversible cessation of spontaneous ventilation and circulation. These changes include: necrosis of brain cells, necrosis of other vital organ cells, cooling, rigor mortis, dependent lividity, and putrefaction. This process actually persists for years, even centuries, until the skeletal remains have disintegrated, and could even be viewed as beginning with the failure of certain organ systems during life. Because these changes occur in a fairly regular and ineluctable fashion, it is claimed that the stipulation of any particular point in this process as the moment of death is arbitrary.

The following argument, however, shows the theoretical inadequacy of any definition which makes death a process. If we regard death as a process, then either (1) the process starts when the person is still living, which confuses the process of death with the process of dying, for we all regard someone who is dying as not yet dead, or (2) the process of death starts when the person is no longer alive, which confuses the process of death with the process of disintegration. Death should be viewed not as a process but as the event that separates the process of dying from the process of disintegration.

On a practical level, regarding death as a process makes it impossible to declare the time of death with any precision. This is not a trivial issue. There are pressing medical, legal, social, and religious reasons to declare the time of death with some precision, including the interpretation of wills, burial times and procedures, mourning times, and decisions regarding the aggressiveness of medical support. There are no countervailing practical or theoretical reasons for regarding death as a process rather than an event in formulating a definition of death. We shall say that death occurs at some definite time, although this time may not always be specifiable with complete precision.

Choices for a definition of death

The definition of death must capture our ordinary use of the term, for *death*, as noted earlier, is a word used by everyone and is not primarily a medical or legal term. In this ordinary use, certain facts are assumed, and we shall assume them as well. Therefore we shall not apply our analysis to science fiction speculations, for example, about brains continuing to function independently of the rest of the organism (Gert, 1967, 1971). Thus we shall assume that all and only living organisms can die, that the living can be distinguished from the dead with very good reliability, and that the moment when an organism leaves the former state and enters the latter can be determined with a fairly high degree of precision. We shall regard death as permanent. We know that some people claim to have been dead for several minutes and then to have returned to life, but we regard this as only a dramatic way of saying that consciousness was temporarily lost (for example, because of a brief episode of cardiac arrest).

Although there are religious theories that death involves the soul leaving the body, we know that religious persons and secularists do not disagree in their ordinary application of the term *dead*. We acknowledge that the body can remain physically intact for some time after death and that some isolated parts of the organism may continue to function (for example, it is commonly believed that hair and nails continue to grow after death). We shall now present our definition of death and contrast it to a proposed alternative.

We define death as the permanent cessation of functioning of the organism as a whole. By the organism as a whole, we do not mean the whole organism, that is, the sum of its tissue and organ parts, but rather the highly complex interaction of its organ subsystems. The organism need not be whole or complete—it may have lost a limb or an organ (such as the spleen)—but it still remains an organism.

By the functioning of the organism as a whole, we mean the spontaneous and innate activities of integration of all or most subsystems (for example, neuroendocrine control) and at least limited response to the environment (for example, temperature change). However, it is not necessary that all of the subsystems be integrated. Individual subsystems may be replaced (for example, by pacemakers, ventilators, or pressors) without changing the status of the organism as a whole.

It is possible for individual subsystems to function for a time after the organism as a whole has permanently ceased to function. Spontaneous ventilation ceases either immediately after or just before the permanent cessation of functioning of the organism as a whole, but spontaneous circulation, with artificial ventilation, may persist for up to two weeks after the organism as a whole has ceased to function.

An example of an activity of the organism as a whole is temperature regulation. The control of this complex process is located in the hypothalamus and is important for normal maintenance of all cellular processes. It is lost when the organism as a whole has ceased to function.

Consciousness and cognition are sufficient to show the functioning of the organism as a whole in higher animals, but they are not necessary. Lower organisms never have consciousness and even when a higher organism is comatose, evidence of the functioning of the organism as a whole may still be evident, for example, in temperature regulation.

We believe that the permanent cessation of the functioning of the organism as a whole is what has traditionally been meant by death. This definition retains death as a biological occurrence which is not unique to human beings; the same definition applies to other higher animals. We believe that death is a biological phenomenon and should apply equally to related species. When we talk of the death of a human being, we mean the same thing as we do when we talk of the death of a dog or a cat. This is supported by our ordinary use of the term *death*, and by law and tradition. It is also in accord with social and religious practices and is not likely to be affected by future changes in technology.

An alternative definition of death as the irreversible loss of that which is essentially significant to the nature of man has been proposed by Veatch (1976). Though this definition initially seems very attractive, it does not state what we ordinarily mean when we speak of death. It is not regarded as self-contradictory to say that a person has lost that which is essentially significant to the nature of man, but is still alive. For example, we all acknowledge that permanently comatose patients in chronic vegetative states are sufficiently brain-damaged that they have irreversibly lost all that is essentially significant to the nature of man but we still consider them to be living (for example, Karen Ann Quinlan; see Beresford, 1977).

The patients described by Brierley and associates (1971) are also in this category. These patients had complete neocortical destruction with preservation of the brainstem and diencephalic (posterior brain) structures. They had isoelectric (flat) electroencephalograms (EEGs) (indicating neocortical death) and were permanently comatose, although they had normal spontaneous breathing and brainstem reflexes; they were essentially in a permanent, severe, chronic vegetative state (Jennett and Plum, 1972). They retained many of the vital functions of the organism as a whole, including neuroendocrine control (that is, homeostatic interrelationships between the brain and various hormonal glands) and spontaneous circulation and breathing.

This alternative definition actually states what it means to cease to be a person rather than what it means for that person to die. *Person* is not a biological concept but rather a concept defined in terms of certain kinds of abilities and qualities of awareness. It is inherently vague. Death is a biological concept. Thus in a literal sense, death can be applied directly only to biological organisms and not to persons. We do not object to the phrase "death of a person," but the phrase in common usage actually means the death of the organism which was the person. For example, one might overhear in the hospital wards, "The person in room 612 died last night." In this common usage, one is referring to the death of the organism which was a person. By our analysis, Veatch (1976) and others have used the phrase "death of a person" metaphorically, applying it to an organism which has ceased to be a person but has not died.

Without question, consciousness and cognition are essential human attributes. If they are lost, life has lost its meaning. A patient in a chronic vegetative state is usually regarded as living in only the most basic biological sense. But it is just this basic biological sense that we want to capture in our definition of death. We must not confuse the death of an organism which was a person with an organism's ceasing to be a person. We are immediately aware of the loss of personhood in these patients and are repulsed by the idea of continuing to treat them as if they were persons. But were we to consider these chronic vegetative patients as actually dead, serious problems would arise. First, a slippery slope condition would be introduced wherein the question could be asked: How much neocortical damage is necessary before we declare a patient dead? Surely patients in a chronic vegetative state, although usually not totally satisfying the tests for neocortical destruction, have permanently lost their consciousness and cognition. Then what about the somewhat less severely brain-damaged patient?

By considering permanent loss of consciousness and cognition as a criterion for ceasing to be a person and not for death of the organism as a whole, the slippery slope phenomenon is put where it belongs: not in the definition of death, but in the determination of possible grounds for nonvoluntary euthanasia, that is, providing possible grounds for killing the organism, or allowing it to die, in those instances in which the organism is no longer a person. The justification of nonvoluntary euthanasia must be kept strictly separate from the definition of death. Most of us would like our organism to die when we cease to be persons, but this should not be accomplished by blurring the distinctions between biological death and the loss of personhood.

When an organism ceases to be a person, that is, when it permanently loses all consciousness and cognition, then practical problems arise. How are we to treat this organism? (1) Should we treat it just as we treat a person, making every effort to keep it alive? (2) Should we cease caring for it, either in part or at all, and allow it to die? (3) Should we kill it?

In our view, an organism that is no longer a person has no claim to be treated as a person. But just as one treats a corpse with respect, even more so would one expect that such a living organism be treated with respect. This does not mean, however, that one should strive to keep the organism alive. No one benefits by doing this; on the contrary, given the care needed to keep such an organism alive, it seems an extravagant waste of both economic and human resources to attempt to do so. On the other hand, it seems unjustified to require anyone to actively kill it. Even though the organism is no longer a person, it still looks like a person, and unless there are overwhelming reasons for killing it, it seems best not to do anything that might weaken the prohibition against killing. This leaves the second alternative, discontinuing all care and allowing the patient to die. This can take either of two forms: discontinuing medical treatment or discontinuing all ordinary and routine care. The latter is the position we favor.

It is important to note that the patient will not suffer from lack of care, for since the patient is no longer a person this means that it has permanently lost all consciousness and cognition. Any patient who retains even the slightest capacity to suffer pain or discomfort of any kind remains a person and must be treated as such. We make this point to emphasize our position that only patients who have completely and permanently lost all consciousness and cognition should have all care discontinued. We believe that discontinuing all care and allowing the patient who is no longer a person to die is the preferred alternative, and the one that should be recommended to the legal guardian or next of kin as the course of action to be followed.

The criterion of death

We have argued that the correct definition of death is permanent cessation of functioning of the organism as a whole. We will now inspect the two competing criteria of death: (1) the permanent loss of cardiopulmonary functioning and (2) the total and irreversible loss of functioning of the whole brain.

Characteristics of optimum criteria and tests

Given that death is the permanent cessation of functioning of the organism as a whole, a criterion will yield a false-positive if it is satisfied, and yet it would still be possible for that organism to function as a whole. By far the most important requirement for a criterion of death is that it yield no false-positives.

A criterion of death, however, cannot have any exceptions; this is what enables it to serve as a legal definition of death. It is not sufficient that the criterion be correct 99.99 percent of the time. This means that not only can the criterion yield no false-positives, it can also yield no false-negatives. A criterion of death yields a false-negative if it is not satisfied and yet the organism as a whole has irreversibly ceased to function. Of course, one may sometimes determine death without using the criterion, but it can never be that the criterion is satisfied and yet the person is not dead, or that the criterion is not satisfied and the person is dead. This is why it is so easy to mistake a criterion for an ordinary definition; it is rather a kind of operational definition, and serves as part of the legal definition, but the real operational definition is provided by the tests which show whether or not the criterion is satisfied.

Permanent loss of cardiopulmonary functioning

Permanent termination of heart and lung function has been used as a criterion of death throughout history. The ancients observed that all other bodily functions ceased shortly after cessation of these vital functions, and the irreversible process of bodily disintegration inevitably followed. Thus permanent loss of spontaneous cardiopulmonary function was found to predict permanent nonfunctioning of the organism as a whole. Further, if there were no permanent loss of spontaneous cardiopulmonary function, then the organism as a whole continued to function.

Therefore permanent loss of cardiopulmonary function served as an adequate criterion of death.

Because of current ventilation/circulation technology, permanent loss of spontaneous cardiopulmonary functioning is no longer necessarily predictive of permanent nonfunctioning of the organism as a whole. Consider a conscious, talking patient who is unable to breathe because of suffering from poliomyelitis and who requires an iron lung (thus having permanent loss of spontaneous pulmonary function), who has also developed asystole (loss of spontaneous heartbeat) requiring a permanent pacemaker (thus having permanent loss of spontaneous cardiac function). It would be absurd to regard such a person as dead.

It might be proposed that it is not the permanent loss of *spontaneous* cardiopulmonary function that is the criterion of death, but rather the permanent loss of all cardiopulmonary function, whether spontaneous or artificially supported. But now that ventilation and circulation can be mechanically maintained, an organism with permanent loss of whole brain functioning can have permanently ceased to function as a whole days to weeks before the heart and lungs cease to function with artificial support. Thus this supposed criterion would not be satisfied, yet the person would be dead. The heart and lungs now seem to have no unique relationship to the functioning of the organism as a whole. Continued artificially supported cardiopulmonary function is no longer perfectly correlated with life, and permanent loss of spontaneous cardiopulmonary functioning is no longer perfectly correlated with death.

Total and irreversible loss of whole brain functioning

The criterion for the cessation of functioning of the organism as a whole is the permanent loss of functioning of the entire brain. This criterion is perfectly correlated with the permanent cessation of functioning of the organism as a whole because it is the brain that is necessary for the functioning of the organism as a whole. It integrates, generates, interrelates, and controls complex bodily activities. A patient on a ventilator with a totally destroyed brain is merely a preparation of artificially maintained subsystems since the organism as a whole has ceased to function.

The brain generates the signal for breathing through brainstem ventilatory centers and aids in the control of circulation through brainstem blood pressure control centers. Destruction of the brain produces apnea (inability to breathe) and generalized vasodilatation (opening of the peripheral blood vessels); in all cases, despite the most aggressive support, the adult heart stops within a week and that of the child within two weeks (Ingvar et al., 1978). Thus when the organism as a whole has ceased to function, the artificially supported vital subsystems quickly fail. Many other functions of the organism as a whole, including neuroendocrine control, temperature control, food-searching behaviors, and sexual activity, reside in the more primitive regions (hypothalamus, brainstem) of the brain. Thus total and irreversible loss of functioning of the whole brain and not merely the neocortex is required as the criterion for the permanent loss of the functioning of the organism as a whole.

Using permanent loss of functioning of the whole brain as the criterion for death of the organism as a whole is also consistent with tradition. Throughout history, whenever a physician was called to ascertain the occurrence of death, his examination included the following important signs indicative of permanent loss of functioning of the whole brain: unresponsivity, lack of spontaneous movements including breathing; and absence of pupillary light response. Only one important sign, lack of heartbeat, was not directly indicative of whole brain destruction. But since the heartbeat stops within several minutes of apnea, permanent absence of the vital signs is an important sign of permanent loss of whole brain functioning. Thus, in an important sense, permanent loss of whole brain functioning has always been the underlying criterion of death.

The tests of death

Given the definition of death as the permanent cessation of functioning of the organism as a whole, and the criterion of death as the total and irreversible cessation of functioning of the whole brain, the next step is the examination of the available tests of death. The tests must be such that they will never yield a false-positive result. Of secondary importance, they should produce few and relatively brief false-negatives.

Cessation of heartbeat and ventilation

The physical findings of permanent absence of heartbeat and respiration are the traditional tests of death. In the vast majority of deaths not complicated by artificial ventilation, these classic tests are still applicable. They show that the criterion of death has been satisfied since they always quickly produce permanent loss of functioning of the whole brain. However, when mechanical ventilation is being used, these tests lose most of their utility due to the production of numerous false-negatives for as long a time as one to two weeks, that is, death of the organism as a whole with still intact circulatory-ventilatory subsystems. Thus though the circulation-ventilation tests will suffice in most instances of death, if there is artificial maintenance of circulation or ventilation the special tests for permanent cessation of whole brain functioning will be needed.

Irreversible cessation of whole brain functioning

Numerous formalized sets of tests have been established to determine that the criterion of permanent loss of whole brain functioning has been met. These include, among others, tests described by the Harvard Medical School Ad Hoc Committee (Beecher, 1968) and the National Institutes of Health Collaborative Study of Cerebral Survival (1977). They have all been recently reviewed (Black, 1978; Molinari, 1978). What we call tests have sometimes been called "criteria," but it is important to distinguish these second-level criteria from the first-level criteria. While the first-level criteria must be for the death of the organism and must be understandable by the layman, the second-level criteria (tests) determine the permanent loss of functioning of the whole brain and need not be understandable by anyone except qualified clinicians. To avoid confusion, we prefer to use the designation "tests" for the second-level criteria.

All the proposed tests require total and permanent absence of all functioning of the brainstem and both hemispheres. They vary slightly from one set to another, but all require unresponsivity (deep coma), absent pupillary light reflexes, apnea (inability to breathe), and absent brainstem reflexes. They also require the absence of drug intoxication and low body temperature, and the newer sets require the demonstration that a lesion of the brain exists. Isoelectric (flat) EEGs are generally required, and tests disclosing the absence of cerebral blood flow are of confirmatory value (NIH Collaborative Study, 1977). All tests require the given loss of function to be present for a particular time interval, which in the case of the absence of cerebral blood flow may be as short as thirty minutes.

Current tests of irreversible loss of whole brain function may produce many false-negatives of a sort during the thirty-minute to twenty-four-hour interval between the successive neurologic examinations which the tests require. Certain sets of tests, particularly those requiring electrocerebral silence by EEG, may produce false-negatives if an EEG artifact is present and cannot confidently be distinguished from brain wave activity. Generally, a few brief false-negatives are tolerable and even inevitable, since tests must be delineated conservatively in order to eliminate any possibility of false-positives.

There are many studies which show perfect correlation between the loss of whole brain function tests of the Ad Hoc Committee of the Harvard Medical School and total brain necrosis at postmortem examination. Veith et al. (1977a) conclude that "the validity of the criteria [tests] must be considered to be established with as much certainty as is possible in biology or medicine" (p. 1652). Thus, when a physician ascertains that a patient satisfies the validated loss of whole brain function tests, he can be confident that the loss of whole brain functioning is permanent. Physicians should apply only tests which have been completely validated.

The determination of death

It is helpful to consider several examples of death to illustrate the applications of our analysis. We will review deaths by primary respiratory arrest, primary cardiac arrest, and primary brain destruction. In each case the process of dying, the event of death, and the process of disintegration will be identified.

Death by hanging

In a properly executed hanging, a displaced fracture of the upper spine ("hangman's fracture") occurs which acutely compresses the cervical spinal cord and produces instantaneous apnea. Some degree of airway compromise and carotid artery compression undoubtedly also occurs, but we will restrict our attention to the principal fatal effect, that of primary respiratory arrest.

As apnea persists, progressive lack of oxygen in the blood and acidosis (body chemistry disturbance) produce cardiac arrest within several minutes. Throughout this time, the brain is becoming progressively less well oxygenated and finally, within a short period after cardiac arrest, suffers total infarction (death from lack of oxygen). Cooling and rigor mortis inevitably follow.

While some might find it attractive to state that death occurs at the moment of the neck fracture, clearly this is not the case. The neck fracture and subsequent respiratory and cardiac arrests are part of the process of dying. Only once the whole brain has permanently lost its function, secondary to the cardiopulmonary arrest, has the event of death occurred. This event is then followed by the process of disintegration. If the victim had been placed on a ventilator immediately after the neck fracture, permanent loss of whole brain functioning would have been prevented. Thus the process of dying would have been reversed and death prevented. A similar and even more dramatic example is that of death by decapitation, though at present there may be no way to reverse the process of dying when the head has been severed from the rest of the body.

Death by chronic disease

Many patients dying of chronic diseases suffer spontaneous ventricular fibrillation (a type of inadequate beating of the heart). As cardiac output ceases, so does cerebral blood flow, so that the brain becomes progressively less well oxygenated. As the brainstem ceases to function, the patient becomes unable to breathe which accelerates the loss of brain and other organ functioning. Finally, the brain becomes totally infarcted, and events of disintegration proceed.

While ventricular fibrillation may be said to be the proximal cause of death, clearly the patient is not dead at the moment the heart stops or even at the time the respirations cease. At these times he is dying. Death occurs when the brain has totally and irreversibly lost its function through ischemic infarction. A timely resuscitation prior to total loss of brain functioning would reverse the process of dying. Death, of course, is irreversible by definition.

Death by massive head injury

These injuries are often complicated by immediate and severe swelling of the brain. When the resulting increased intracranial pressure exceeds that of systolic blood pressure, cerebral blood flow ceases and whole brain infarction occurs. At the time of head injury or minutes later, while the whole brain is dying, brainstem failure causes breathing to stop. Progressive loss of blood oxygenation and acidosis (body chemistry disturbance) then produce an inadequate or absent beating of the heart. Untreated, the organism undergoes cooling and rigor mortis.

However, if the patient were placed on a ventilator immediately after irreversible loss of whole brain function, but prior to cardiac arrest, the circulatory functioning could probably be maintained for a few days. If bedside testing confirmed satisfaction of validated loss of whole brain function tests, the difficult situation would be present of preserved circulatory-ventilatory subsystems in an organism which had ceased to function as a whole. Death would have occurred at the time the brain had totally and permanently lost all functioning. In this case, mechanical ventilation and other aggressive therapeutics would have delayed the process of disintegration, although the event of death had already occurred.

The statute of death

In July, 1981, the President's Commission for the Study of Ethical Problems in Medicine and Biomedical and Behavioral Research published its report, "'Defining Death,' a Report on the Medical, Legal, and Ethical Issues in the Determination of Death."

In recommending a statutory definition of death, they attempted to reconcile conceptual and theoretical considerations on the one hand and practical considerations on the other. We believe that conceptual and theoretical considerations should lead one to propose a statute which uses irreversible cessation of the entire brain, including the brainstem, as the criterion (standard) of death. (The Commission uses the term "standard" as we use the term "criterion.") Practical considerations should lead one to propose a statute which includes prolonged loss of spontaneous circulatory and respiratory functions as a test of death. The goal of a statute is to be conceptually and theoretically clear while at the same time responding to what the Commission regards as the practical purpose of guiding "the beliefs and behavior of physicians and the public." We do not think that the Uniform Determination of Death Act (UDDA) statute which the Commission recommends adequately meets this goal. The UDDA provides:

An individual who has sustained either (1) irreversible cessation of circulatory and respiratory functions, or (2) irreversible cessation of all functions of the entire brain, including the brainstem, is dead. A determination of death must be made in accordance with accepted medical standards.

A criterion (standard) of death is not merely that by which we can recognize that someone is dead; it is, based on all of our medical understanding, that which is both a necessary and sufficient condition for death. If the criterion (standard) is fulfilled, the person is dead; if it is not fulfilled, the person is not dead. Irreversible cessation of all brain functions is such a criterion (standard). If it has occurred, the person is dead; if it has not occurred, the person is not dead, no matter what has happened to the heart, lungs, or anything else. It is irrelevant that one can recognize that someone is dead without thinking about the brain, for whether he thinks about it or not, we only accept that recognition as correct because we know that, for example, prolonged cessation of heart and lung function is perfectly correlated with, and thus an adequate test of, the brain criterion (standard) being fulfilled.

Let us now see why "irreversible cessation of circulatory and respiratory functions" is not an acceptable criterion (standard) of death. First, we must note that there is an ambiguity in this phrase. (Recall the parallel problem with "going around." See Chapter 1, pp. 14ff.) This phrase can mean either "irreversible cessation of *spontaneous* circulatory and respiratory functions," or it may mean "irreversible cessation of *artificially supported* circulatory and respiratory functions." There is no such ambiguity with regard to cessation of brain functions, for there are no artificially supported brain functions, in the relevant sense. Irreversible cessation of spontaneous circulatory and respiratory function is not a standard of death, for, as we pointed out earlier (p. 186), no one would want to call someone in an iron lung and wearing a pacemaker dead, especially if he were still talking to us. Thus, though irreversible cessation of spontaneous heart and lung function may be a necessary condition of death, it is certainly not sufficient. Irreversible cessation of artificially supported circulatory and respiratory functions is also not a criterion (standard) of death; though it may be a sufficient condition of death, it is not a necessary condition, for, as the Commission agrees, someone whose circulation and respiration are being artificially maintained, but all of whose brain functions have irreversibly ceased, is dead.

One might try to interpret the phrase as "irreversible loss of *either spontaneous or artificially supported* circulatory and respiratory functions," but this will not work either, for this entails that irreversible loss of spontaneous circulatory and respiratory functions would be sufficient for death. Nor will interpreting the phrase as "irreversible loss of *both spontaneous and artificially supported* circulatory and respiratory functions" do, for this entails that irreversible loss of artificially supported circulatory and respiratory functions is a necessary condition of death. We conclude that there is no interpretation of the phrase "irreversible loss of circulatory and respiratory functions" such that it provides an acceptable criterion (standard) for death.

No doubt some of the Commission's problems were brought on by their concern with avoiding radical change. The following passage indicates their view:

The conservative nature of the reform here proposed will be more apparent if the statute refers explicitly to the existing cardiopulmonary standard for determination of death. The brain-based standard is, after all, merely supplementary to the older standard, which will continue to be adequate in the overwhelming number of cases in the foreseeable future. (p. 63)

What they do not seem to realize is that any criterion (standard) of death must be adequate in all cases, not merely the overwhelming number of cases. Here is another instance where the term "spontaneous" becomes important. For the "older standard" was for the prolonged absence of *spontaneous* circulatory and respiratory functions, and the Commission rightly recognizes that this standard is no longer universally adequate. What they do not seem to realize is that this means it is not a standard but merely a test. It may be that the cardiopulmonary tests will be adequate in the overwhelming number of cases, and that the brain-based tests will be used in only a small proportion of cases, but this belongs to the practical part of the statute, not in the statutory definition of death.

Given our new medical understanding, it is clear that only the irreversible cessation of all brain functions can count as a standard of death. Though practical considerations may lead one to include prolonged loss of spontaneous circulatory and respiratory functions in a statute, such practical considerations should not be allowed to overrule conceptual and theoretical considerations. By allowing this, the Commission has recommended a statutory definition of death which is seriously misleading, and which has the most serious flaw that the Commission finds in previous statutes: providing two independent criteria (standards) of death, without explaining the relationship between them. (See for example, Capron and Kass, 1972).

A conceptually satisfactory statute would not need to mention cessation of cardiopulmonary function at all. It would be sufficient to include only irreversible cessation of whole brain functioning and allow physicians to select validated and agreed upon tests (of which prolonged absence of spontaneous cardiopulmonary function would be one) to measure irreversible cessation of whole brain function. However, we agree with the Commission that a statute would be more broadly acceptable and useful if cessation of cardiopulmonary function were included in it, but we maintain that the role of cessation of spontaneous cardiopulmonary function is a test for death, not a criterion (standard) of death.

The solution to statute design, then, is to reconcile the claims of conceptual clarity and practical considerations. In order to produce a more conceptually acceptable statute of death, we have incorporated into the UDDA statute the distinction between a criterion (standard) and a test that was recognized by the model statute provided by the Law Reform Commission of Canada, and by the statute we recently proposed (Bernat, Culver, and Gert, 1981). It reads:

An individual who has sustained irreversible cessation of all functions of the entire brain, including the brainstem, is dead

(a) In the absence of artificial means of cardiopulmonary support, death (the irreversible cessation of all brain functions) can be determined by the prolonged absence of spontaneous circulatory and respiratory functions.

(b) In the presence of artifical means of cardiopulmonary support, death (the irreversible cessation of all brain functions) must be determined by tests of brain function.

In both situations, the determination of death must be made in accordance with accepted medical standards.

We believe that this statute is conceptually clearer than the UDDA statute. It identifies the irreversible cessation of all functions of the entire brain, including the brainstem, as the standard of death, thus making clear that death is a single phenomenon. We believe that it also provides a practical guide to the practicing physician, pointing out that he may continue to declare death by cardiopulmonary tests in the majority of deaths uncomplicated by artificial cardiopulmonary support. And in the presence of cardiopulmonary support, he must directly measure the functioning of the brain. Thus it is incremental in its practical function, while maintaining conceptual clarity.

By using the prolonged absence of spontaneous circulatory and respiratory functions as a test for irreversible loss of whole brain function, our proposed statute allows us to answer the question raised by Jonas (1974): "Why are they alive if the heart, etc., works naturally but not alive when it works artificially?" In our account, spontaneous circulation and ventilation show that at least part of the brain continues to function, whereas artificially supported circulation and ventilation does not show this. Thus in the latter case one must discover directly if the whole brain has permanently ceased to function.

Note

1. This chapter is adapted in part from Bernat, Culver, and Gert (1981).

References

American Bar Association. House of Delegates redefines death, urges redefinition of rape, and undoes the Houston amendments. *American Bar Association Journal*, 1975, *61*, 463–464.

Beecher, Henry K. A definition of irreversible coma: report of the Ad Hoc Committee of the Harvard Medical School to examine the definition of brain death. *Journal of the American Medical Association*, 1968, *205*, 337–340.

Beresford, H. Richard. The Quinlan decision: problems and legislative alternatives. *Annals of Neurology*, 1977, *2*, 74–81.

Bernat, James L., Culver, Charles M., and Gert, Bernard. On the definition and criterion of death. *Annals of Internal Medicine*, 1981, *94*, 389–394.

Black, Peter M. Brain death. *New England Journal of Medicine*, 1978, *299*, 338–344, 393–401.

Brierley, J. B., Adams, J. H., Graham, D. I., and Simpson, J. A. Neocortical death after cardiac arrest. *Lancet*, 1971, *2*, 560–565.

Capron, Alexander M., and Kass, Leon R. A statutory definition of the standards for determining human death: an appraisal and a proposal. *University of Pennsylvania Law Review*, 1972, *121*, 87–118.

Gert, Bernard. Can the brain have a pain? *Philosophy and Phenomenological Research*, 1967, *27*, 432–436.

Gert, Bernard. Personal identity and the body. *Dialogue*, 1971, *10*, 458–478.

Hastings Center Task Force on Death and Dying. Refinements in criteria for the determination of death: an appraisal. *Journal of the American Medical Association*, 1972, *221*, 48–53.

Ingvar, David H., Brun, Arne, Johansson, Lars, and Sammuelsson, Sven M. Survival after severe cerebral anoxia with destruction of the cerebral cortex: the apallic syndrome. *Annals of the New York Academy of Science*, 1978, *315*, 184–214.

Jennett, B., and Plum, F. Persistent vegetative state after brain damage. A syndrome in search of a name. *Lancet*, 1972, *1*, 734–737.

Jonas, Hans. *Philosophical Essays: From Ancient Creed to Technological Man.* Englewood Cliffs, N.J.: Prentice-Hall, 1974, pp. 134–140.

Law Reform Commission of Canada. *Criteria for the Determination of Death.* Ottawa: Law Reform Commission of Canada, 1979.

Molinari, Gaetano F. Review of clinical criteria of brain death. *Annals of the New York Academy of Science*, 1978, *315*, 62–69.

Morison, Robert S. Death: process or event? *Science*, 1971, *173*, 694–698.

NIH Collaborative Study of Cerebral Survival. An appraisal of the criteria of cerebral death: a summary statement. *Journal of the American Medical Association*, 1977, *237*, 982–986.

President's Commission for the Study of Ethical Problems in Medicine and Biomedical and Behavioral Research. *"Defining Death," a Report on the*

Medical, Legal and Ethical Issues in the Determination of Death. Washington, D.C., 1981.

Veatch, Robert M. *Death, Dying and the Biological Revolution: Our Last Quest for Responsibility.* New Haven, Conn.: Yale University Press, 1976.

Veith, Frank J., Fein, Jack M., Tendler, Moses D., Veatch, Robert M., Kleiman, Marc A., and Kalkines, George. Brain death I. A status report of medical and ethical considerations. *Journal of the American Medical Association,* 1977a, *238,* 1651–1655.

Veith, Frank J., Fein, Jack M., Tendler, Moses D., Veatch, Robert M., Kleiman, Marc A., and Kalkines, George. Brain death II. A status report of legal considerations. *Journal of the American Medical Association,* 1977b, *238,* 1744–1748.

Index